Exhibition Poultry Breeder's Handbook

The author with a student Troy O'Connell and a Light Sussex female. Photograph courtesy *The Land*.

Exhibition Poultry Breeder's Handbook

Rick Kemp

Kangaroo Press

Dedication

To the Poultry Fancy and all those who enjoy pure breed poultry, and

To the memory of James Hadlington, former N.S.W. Department of Agriculture poultry expert, whose work was regarded as the 'by word' on poultry during the era when pure breed poultry formed the basis of the poultry industry in Australia. It was largely through the work of Hadlington that Australia's national commercial breed, the Australorp was nationally standardised.

First published in 1989 by Kangaroo Press Pty Ltd
Second edition by Kangaroo Press in 1997
3 Whitehall Road Kenthurst NSW 2156
PO Box 6125 Dural Delivery Centre NSW 2158
Printed by Kyodo Printing Co. (S'pore) Pte Ltd

ISBN 0 86417 835 2

Contents

Acknowledgments

The author wishes to express his gratitude to the following:

My wife, Sandra and family, for the time they have been deprived of whilst putting this book together.

Mr Bert E. Brown for poultry resource material.

Mr Bob Venn for poultry resource material.

Mr Bill Bruce for photographs and poultry history.

Mr Allan Sharpe for poultry companionship and a wealth of specific breed information.

The *Land* newspaper for photographs of self.

Peter and Helen Gay for photographs and information on autosexing breeds.

Mrs Evelyn Chalmers for Legbar photographs and information.

The following people who made it possible to photograph selected fowls at the 1988 Sydney Royal Easter Show

Mr David White, R.A.S. councillor and Steward in Chief, Poultry Section.

Mr Peter Smith, R.A.S. Assistance Steward in Chief.

Mr Jeff Thompson, R.A.S. Poultry Pavilion Supervisor.

Mr Kerry Pearce, R.A.S. Clerk of the Pavilion.

Mr Barry Reeves, Steward.

Mr Brian Weis, Steward.

Mr B. Watling, President and Mr B. Mason, Secretary of the Fairfield Poultry Club and Rare Breeds Society for permission to photograph at their 1988 Autumn Show.

Mr G. Childs, President and Mr J. List, Secretary of the Camden Poultry Club for permission to photograph at their 1988 Annual Show.

Mr L. Sinclair, President and Mr R. Harris of the Bantam Club of N.S.W. for permission to photograph at their 1988 Bicentennial Show.

Mr D. Plant, Secretary of the Pekin Club of Australia for permission to photograph at their 1988 Annual Show.

Mr M. O'Connor, Secretary of the Ancona Club of Australia for permission to photograph at their 1988 Annual Show.

Mr G. Sharpe, Secretary of the Leghorn Club of Australia for permission to photograph at their 1988 Annual Show.

The following people who made their fowls available for photographing to illustrate this book.

S. Adams	R. Forrest	M. O'Rourke
B. & E. Bell	S. Fraser	M. Pearson
K. Bergin	G. Frazer	R. & J. Perkinson
B. Boardman	J. Gardner	B. Raines
B. Bulter	E. Green	J. Rogers
B. Burnham	F. Greenaway	D. Roth
F. Catt	C. Gregor	M. Schultz
A. Cheetham	R. Guy	G. Sharpe
G. Childs	K. Heydon	G. Sheaves
P. Chowne	H. Holyoake	J. Sibley
J. Clark	L. Huntingdon	L. Simone
D. Clarke-Bruce	L.K. King	L. Sinclair
L. Condon	K. & G. Lambert	A. & M. Smith
K. Cook	A. Lamont	P. Smith
A. Cooksley	G. Lee	C. Ubrihien & Sons
L. Coombes	D. & J. Leyshon	R. Thelan
G. Coote	S. & D. Lindsay	M. & D. Thompson
J. Crabb	J. List	M. Thompson
R. Cupitt	M. Lye	R. Towner
T. & L. Davis	B. McCredie	B. Trimmer
W.K. Dickson	C. McKenzie	B. Usope
K.R. Dubber	D.H. McKenzie	B. Weiss
B. Dukes	N. Middlebrook	C. White
M. Filmer	A. & S. Mills	D. White
E. Flarrety	O. Mills	W. Wingett
B. Ford	D. Newman	J. Wong
F. Fogarty	S. & A. Nicholas	G. Wood
L.T. Ford	M. O'Connor	

Hills District Personnel for typing and manuscript preparation.

Introduction

Welcome. This book is to help people interested in pure breeds of poultry become genuine breeders of Standard bred poultry. From the outset, it must be pointed out that this book does not set out to replace the various breed Standards by which poultry are judged. More the book should be seen as an adjunct to help people achieve the various Standards laid down. It would therefore be in the interests of the reader to purchase a copy of the Standard whether it be The British Poultry Standards, or American Standard of Perfection, or American Bantam Standard or to join the appropriate breed club to obtain the necessary Standard that governs your chosen breed(s). This then gives a person the final word on the breed and should be the underlying philosophy of every breeding program.

At the present time, what the poultry fancy needs is more dedicated breeders to take on a breed, look after it, foster it and develop lines or strains to ensure the long term survival of that breed. The product will reflect the person's skill as a breeder. To do this job well, the breeder must be prepared to give a breed a fair percentage of the space available in their yards, as many birds have to be kept to build up the necessary nucleus of stock needed to maintain a breed. Far too many so called breeders have a 'poultry show' in their yards with a few birds of many breeds, but not enough to build up the required nucleus of stock. These people fail to see the social benefit of concentrating their efforts into a small top class team, going to a show, looking at the handywork of others, talking to others and encouraging their efforts.

It is unfortunate that each breed of poultry has all too few genuine breed specialists, people who have taken the time to study carefully the 'ins' and 'outs' of a breed. This can only be achieved through a carefully run breeding program over many years and as much reading about the breed as possible. There are plenty of poultry 'showmen' around relying on the annual purchase of a 'good one' from an established breed specialist to lead their show team as they do their round of shows. What happens to the breed when the breed specialist passes on or retires? All the knowledge and breeding stock goes, quite often into the wrong hands, and is lost forever. A total breeding program becomes extinct.

It could become necessary to set up Rare Breeds Trust farms in Australia similar to those seen in Great Britain to ensure the survival of some breeds. Apart from an aesthetic point of view, breeds which currently seem commercially useless may contain genes of importance in future breeding programs. An interesting recent example of this was the discovery that if leg length on commercial laying hens was reduced, the number of eggs damaged during laying would be reduced. It was found that the Scots Dumpy possessed the necessary genes to do this but was unable to be obtained by researchers in its land of origin. In fact, the search for the breed ended in Kenya where a keen breeder had a flock of several hundred. Members of this flock were returned to Great Britain to the Rare Breeds Trust to build up a new flock in the country of its origin! The famous poultry writer W. Powell-Owen once stated, 'Every breed, popular or otherwise, may have some future part to play in the making of other breeds and the advancement of our knowledge of matings, hence the importance of their maintenance'. This should sum up our attitude to all breeds irrespective of our personal opinions of them.

The maintenance of our breeds of poultry should be a general aim of the poultry fancy at this time as we reflect on our heritage over the Bicentennial. Fanciers would be doing their part in the effective celebration of this event if they supported a breed of poultry which has been part of the fabric of this country. Such support may be indirect by encouraging those who are prepared to give the time, space and effort to a breed's long term survival. This encouragement could take the form of interest in what the person is doing, giving honest points of view when asked for an opinion or moral support when the breeder is confronted by knockers.

This book has endeavoured to take a consistent approach when dealing with each breed by grouping the written material under breed history, positive features to look for, negative features to avoid and breeding hints. Some breeds are covered in more detail than others. This is unavoidable because of the varying amounts of material available. The material is based on the author's own experience, research and discussions with many poultry breeders past and present. Accompanying the written material is a pictorial approach to each breed type aimed at building up a mind's eye picture of what each breed should look like. There are some aspects of type which cannot be depicted by photograph or drawing them and these can only be learned by hands on experience with fowls in the yard or pen.

Mention should be made of Breed Standards in general. These are written statements of what a breed ideally should be. They are drawn up by the founders of a breed for acceptance by the body that controls Breed Standards. Once the breed is accepted, a written statement of what the breed should be and look like is available for breeders, judges and others interested. It is on this statement that the breed is assessed in the show pen, and it should be used by breeders in directing their breeding programs as it contains faults that need to be culled. Wherever there is interpretation of the written word, there will be variations in how birds will be assessed. However it is important to regularly read and reread the appropriate Standard as a whole to avoid undue emphasis on one part of it leading to 'faddism', which has destroyed the intentions of many Standard writers and caused breeds to become distorted. If there is one criticism that could be levelled at Standards in general it is that they often fail to use an objective approach, leaving much of what is written open to a wide range of interpretation. This is an area that breed clubs need to pay a lot more attention to.

Some remarks must be made about Game fowls and in particular to the practice of dubbing the males for exhibition. Dubbing refers to the surgical removal of ear lobes, wattles and trimming of combs before they are acceptable for exhibition. If you do not approve of this practice, then do not breed Game fowls that need dubbing. There are plenty of colourful softfeather breeds that offer the colour of Game fowls and that can be enjoyed without the dubbing. It is the author's opinion that it would be very difficult to convince the general public at large that this practice does not cause stress to the bird, and should a public debate be held on the issue Game fanciers would probably come off second best. It is an issue that will have to be faced by Game fanciers. It is not a matter of 'if' it happens but 'when', and they will have to mount a very convincing case or risk being publicly discredited.

The range of breeds available in pure breed poultry offers something that should appeal to everybody, from a small suburban intensive set up using a docile bantam breed to the hobby farmer with ample space to allow a breed to range. It is under such conditions that many of the larger heavy breeds can be raised to display them at their best. If breeds laying different coloured eggs interest, colours ranging from white and tinted to brown, dark brown, blue, green and speckled are available to satisfy those needs. Feather colour ranges from a single colour to intricate patterns challenging enough for even the most ardent breeder. There is something to appeal to everyone.

Once a breed is selected, make sure a reference library of material about that breed is built up. This may take the form of specialist books, photographs, paper clippings and notes made following visits or phone calls to other breed specialists. Such a reference library gives a breeder the deeper understanding of a breed necessary for success. This is often a painstaking process, sometimes with apparent dead-ends in investigations, but it is important to approach it positively. Your keenness reflects well on you.

Should this book create differences of opinion, and no doubt it will, so much the better, as healthy constructive debate leads to a better understanding which must be in the long term interest of the breed.

In conclusion, a quote from the late Fred Owen-French published in the *Poultry* newspaper of 30 August 1962 which sums up the situation of the poultry breeder.

> Breed to the Standard, show to the Standard, judge to the Standard and you will be a 'law abiding' fancier, happier knowing that you have done a good job.

Best of luck.

Breeding

Before looking at individual breeds of fowls, some general remarks on breeding need to be made. The best breeding plans cannot lead to success unless they are integrated with good stockmanship which includes feeding, housing, health and general management. Without proper attention to stockmanship, the bird's true genetic potential will never be realised. Conversely, the best stockmanship will not improve the genetic potential of poorly bred birds. Let's look briefly at each of the above-mentioned points beginning with housing. The prime objective is to protect the birds from extremes of climate, wind, rain or temperature extremes, thus avoiding unnecessary stress. To this end, some breeds may need to be intensively housed to protect their leg plumage or colour, remaining under cover at all times.

A preventative health program is essential to head off potential problems before they take hold in a flock. This would include vaccination for key diseases such as Marek's disease, fowl pox, infectious laryngeal tracheitis (I.L.T.) and infectious bronchitis (I.B.), use of medicated food with young growing stock to build up their resistance to coccidiosis, regular worming treatment and an external parasite reduction program for lice and mites.

It is often quoted that 'the birds are what they are fed', meaning that proper attention must be directed to feeding fowls so that their basic nutritional needs are met. These needs are energy, protein, minerals, vitamins and water. How these are met is up to the individual but perhaps the easiest way to ensure these needs are met is the use of commercially prepared rations supplemented with grain, greenfeed and shellgrit to provide some variety to the bird's diet. Some breeders have favoured feeding regimes but the end result of these is the same, that is, meeting the bird's nutritional needs.

Finally, under the heading of stockmanship is general management which involves the regular 'rounds' of the poultry yard and looking at the fowls, watching for problems that could arise. This involves observation of individual fowls, groups of fowls and handling fowls. In conclusion, good stockmanship

must go hand in hand with good breeding to produce worthwhile results.

The field of poultry genetics itself is very complex. This book does not attempt to offer specific models of the various features of each breed, rather it looks at various phenotypic features (outward appearance) that are part of breeding each breed. For those who wish to seek out genetic model explanations of the inheritance of the various features, the purchase of one of the standard texts in this area would prove a worthwhile investment as a reference book.

Parts of a fowl referred to in the text are summarised in the diagram 'Parts of a Fowl'. A thorough knowledge of these will assist in an understanding of the features to look for and avoid when selecting breeding pens or culling stock.

To conduct a productive breeding program a system of identifying stock and written records are essential. From the time eggs are collected from the breeding pen, through incubation, growing and on to adulthood, every individual should be identifiable. At the beginning of such a system the parent stock must have some form of identification, the best being numbered, coloured leg bands. The breeding pen also needs to be coded and written down in a set of records. Serious breeders use a single mating system to ensure they keep

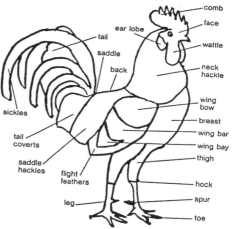

External features of the fowl

track of each mating and in turn, each egg that is laid by a breeding female. Each egg needs to be clearly marked with a pen code or the female's leg band number, and the date it was laid. If an incubator is used, separate divisions can be made in the hatching tray to accommodate eggs from each breeding pen. Upon hatching and before placing in the brooder, the chickens from each female can be web (toe) punched with the breeding pen's assigned marking. This then remains with the chicken for the rest of its life, with its ancestry being easily traced by checking its web (toe) punch code against the breeding records. After the usual culling of young stock, those that remain can be given a numbered leg band which they keep for the rest of their life or until sold or otherwise disposed of. A detailed example of such a breeding record has been included. It is based on a single mating system and using a Multiplo E2 incubator, which allows eggs to be set on a weekly basis. The setting tray is moved down the incubator each week until the eggs are ready to be placed in the hatching tray after the 18th day. The eggs from each mating are then placed in a separate compartment in the hatching tray thus keeping chickens from each mating separated until toe punched.

The hatching tray showing the wire divisions to keep the eggs of different breeding pens separated. The resulting chickens can then be web (toe) punched with the assigned marking for that mating before the chickens are placed in the brooder.

Whilst this is not the only way of designing a breeding record, it does have the following advantages

1. When eggs are collected from the breeding pen, the egg code number and the date laid are written on the egg. An HB grey lead pencil is suitable for this task.

2. The male and female parents are clearly identified by their coloured and numbered leg bands eg R36 for Red 36 and Y6 for Yellow 6.

3. All chickens produced from a mating are given the same toe (web) punch to make them readily identifiable for the rest of their life or stay in your yard.

4. The fertility level is easily checked as the number of eggs set each week from each breeding pen is recorded, the number of fertile eggs placed in the hatching tray after the 18th day is recorded and the number of chickens hatched from the eggs is recorded. As a result, fertility and hatchability of a mating can be easily checked.

5. The number of chickens of either sex is also recorded. This is also important in assessing the rearability of the resulting chickens.

6. Evaluation comments can be made about the mating at the end of the breeding season or year after the progeny have been raised. This information is useful when deciding if the mating is worth continuing with, particularly when tracing inherited faults or desired features.

Most importantly, every serious breeder should know the parentage of every fowl in the yard. This information is vital to the structuring of future breeding programs.

Breeding Strategies

Breeding strategies used in poultry breeding in general can be summarised as follows:

Multiplo E2 incubator showing the three levels of hatching trays permitting eggs to be set on a weekly basis. Each week the setting tray is moved down until on the eighteenth day, the eggs are candled and placed in the hatching tray at the bottom of the incubator.

1. Crossbreeding
2. Inbreeding and Line breeding
3. Criss cross breeding
4. Outbreeding
5. Cockerel and pullet breeding pens

1. Crossbreeding

Crossbreeding, as the name implies, refers to the mating of two different breeds. This is done commecially to utilise the hybrid vigour and productive advantages shown in the progeny, for example, the Australorp—White Leghorn cross used to produce the commercial layer. It would appear that this strategy would have little place in pure breed poultry. However, there are exceptional circumstances when it is used. The most frequent use is when a breed is in a very poor state, generally through prolonged line breeding, or is few in numbers; then the only course of action is to rejuvenate the breed by crossing with another. The breed selected for such a cross is generally one that was used in the breed's original make up prior to standardisation, or a breed similar in type with features that could strengthen the target breed. Examples could include increasing the size of the original small Italian Leghorn with use of Minorcas, and the tightening up of the 'blousey' feathering in Australorps with the use of selected Chinese Langshans. Cross breeding as a course of action must not be taken lightly as this is a long term strategy requiring hatching of large numbers of chickens and drastic culling to achieve the desired individuals in the offspring. Even after stabilisation of the breed once again, periodic faults associated with the infused breed will surface in the offspring. For the average breeder, most breed faults can be rectified by using an unrelated member of the same breed, followed by a big hatch and a severe culling before returning to a line breeding strategy. Such a tactic is often referred to as outcrossing.

2. Inbreeding and Line Breeding

Much discussion could be had over the distinction between the two strategies, as there is a deal of overlap between the two. In general, inbreeding is regarded as the mating of two very closely related fowls, for instance, father to daughter, son to mother, and even in some cases, brother to sister. Line breeding refers to the breeding of fowls from the same general family and whose parentage can be traced back to some reasonably recent common ancestor(s). In other words, they are family related. The main reason why most pure breed fanciers practise inbreeding or line breeding is that it fixes many of the desired features into a line or family of fowls. This process eliminates much of the genetic variability that occurs when unrelated fowls are mated. Whilst this would seem to be the strategy

for consistently producing offspring with the desired breed features, there are some inherent dangers. The fowls selected for a breeding program must be near to perfect for that breed, as not only are the desired features locked into a line, but so too are the undesirable ones. Added to this, if both parents possess a similar fault, this will appear in most of the offspring produced and will only be eliminated by a suitable cross to an unrelated bird.

As well as maintaining accurate breeding records, accurate line or family charts must be kept. On this chart, any highly desirable features should be marked down as well as any undesirable fault(s) so that the faults are not mated in a breeding program. Such a chart has the benefit of locating reasonably closely related birds which could be used in a breeding program, especially to freshen up a stale breeding line without having to resort to a drastic outcross and probable introduction of new undesirable features. A line or family chart has been included to illustrate such a record.

Fairly close line breeding can be undertaken with most breeds for three or four generations but then a family outcross is generally needed to freshen up the vigour of the line. There are some exceptions, generally amongst the Game breeds. The problem with constant inbreeding is that the genes which give resistance to environmental factors are often lost because of the narrowing of the genotype or genetic makeup.

One danger with line breeding is that it can become a mechanical process with some breeders talking in terms of fractions or percentages of certain parents or 'blood'. It is important that a breeder's flare and stockmanship should take precedence over the mechanical approach. Attention must be paid to the heritability of various desired and undesired features as some lines may pass the desired features on more readily than others. The breeder must be attuned to these.

Where utility poultry breeds are involved, many productive features are controlled by groups of genes (polygenic inheritance). It is commonly held that such features, eg egg laying ability, respond more favourably to line breeding than other techniques.

3. Criss Cross Breeding

This can be used in conjunction with a line breeding program and aims to maintain the benefits of line breeding, that is, the consistent production of highly desirable offspring and the benefits of a mild outcross to maintain vigour. As the accompanying illustration shows, a single female or a number of females can be used. The two males used can be brothers if a tight criss cross program is being used or father and son or two line related males selected from a different part

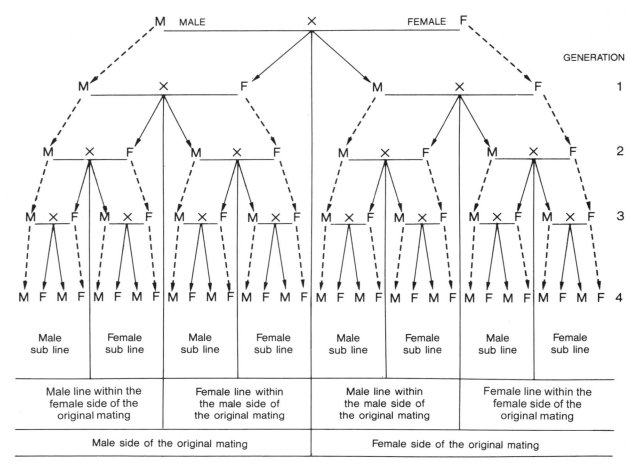

Male sub line	Female sub line	Male sub line	Female sub line	Male sub line	Female sub line	Male sub line	Female sub line
Male line within the female side of the original mating		Female line within the male side of the original mating		Male line within the male side of the original mating		Female line within the female side of the original mating	
Male side of the original mating				Female side of the original mating			

1. This is a line breeding diagram based on using a selected single pair.
2. The outside sub line will give the greatest percentage of the original parent.
3. Sub lines within each side of the mating will give sufficient separation to maintain vigour but are strongly line linked to the desired original parent.
4. Some sub lines may be better producers of good fowls. Using such a chart will increase the chances of isolating the more desired breeding stock.

How to line breed: This illustration is known as Anderson's line-breeding chart and shows that a son may be mated to mother and father to daughters, and how subsequent matings may be made. The black circle represents the male, and the white the female. The chart is read from the top downwards commencing with an unrelated male and female. These mated together produce progeny generally recognised as half blood of each parent. In subsequent matings as outlined on the chart, each solid line represents the male side of the mating and the dotted lines the female. The proportion of male to female blood in the resultant progeny of each mating is indicated by the amount of black and white in each circle. The most essential point in line-breeding is never to breed from any fowl with any defect or evidence of ill-health. It should be remembered that each individual inherits all the characters, both good and bad, of the race, thus the danger of any system of inbreeding of which line-breeding is a modified form, is apparent. The probability of reproducing or intensifying undesirable characters is lessened, however, if the cardinal principle of selecting breeding stock on the basis of stamina is rightly observed.

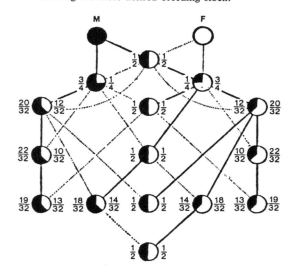

Breeding record

EGG CODE	BREEDING PEN MALE ID	FEMALE ID	CHICKEN TOE PUNCH	EGGS SET 13/9	20/9	27/9	4/10	2/10	HATCHING TRAY 2/10	9/10	16/10	23/10	31/10	CHICKENS HATCHED 4/10	9/10	19/10	25/10	2/11	CHICKENS REARED M	F	COMMENTS eg CULL % Pedigrees
1	Blue Red Dark Leg R36	Blue Red Dark Leg R37	ΨΨ	2	4	4	2		0	2	3	2	2	0	2	0	2	2	3	2	
2	Blue Red Dark Leg R19	Blue Red Dark Leg R20	ΨΨ	1	3	2	2	3	0	1	2	2	2	0	1	1	2	1	2	3	
3	Partridge bred Blue Red Dark Leg Y6	Blue Red bred Partridge Light Leg. Y7	ΨΨ			1													1			Culled poor layer
4	Blue Red Yellow Leg P12	Blue Red L.L. bred ? (Pile) P13	ΨΨ	4	3	4	3	0	2	3	4	3		2	3	4	3		6	6	Discontinue – poor type
5	Blue Red L.L. B9	Top Blue Red. L.L. Hen. B10	ΨΨ	4	1	2	4		4	1	0	2	2	4	0	0	0	1	3	2	
6	Blue Red L.L. R72	Blue Red L.L. Hen No.2. R77	ΨΨ	3	3	2	5	0	3	3	2	5		3?	3	2	4		7	5	
7	Blue Red L.L. P26	B.V.(B.T.W.) Hen. Y8	ΨΨ	3	4	4	3	3	2	4	3	3	2	1	1	0	2	2	2	4	Small weak chickens – outcross next year
8	Black Red LL Ckl. R56	Wheaten Hen. Y49	ΨΨ	4	3				4	3	0	0		4	3	0	0		4	3	
9	Black Red LL Cock	Wheaten Black Tail (B.T.W.) bred.	ΨΨ			1	2			1	2			0	0	0	0	1	1		Ate early eggs

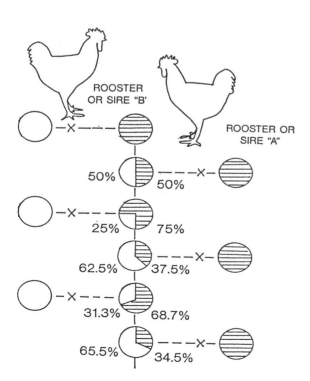

ROOSTER OR SIRE "B"

ROOSTER OR SIRE "A"

50% — 50%

25% — 75%

62.5% — 37.5%

31.3% — 68.7%

65.5% — 34.5%

A plan of criss cross mating showing the proportion of sire "A" and sire "B" in successive generations. Using two unrelated sires, vitality in a line of fowls can be maintained. If a tight line is favoured, two closely related males can be used.

of a line or family chart. The percentage of each male's influence is also given so if one male is preferred to the other, an extra cross to that male will yield a greater proportion of his influence. This strategy can be used in the 'cockerel' and 'pullet' breeding situation as well as the 'single pen breeds'.

4. Outbreeding

When a line breeding program appears to have run into a 'dead end', that is, not producing any worthwhile offspring because of loss of vigour or too many faults appearing, it may be necessary to introduce a new unrelated sire or dam from another breeder. There is much discussion as to whether a male or female should be used in this case. Most say that a female should be introduced and her offspring worked back into the original line or family. Specialist breeders are reluctant to take this step as there is no guarantee that the two blood lines will improve the situation and produce the desired offspring, or 'nick' or 'click' as breeders say. One line may dominate the other to the offspring's disadvantage. Because there is a new assortment of genes, the offspring often do not appear uniform, reflecting the new gene assortment. Large numbers may have to be bred and a severe culling program used to isolate the desired individuals. Sometimes the whole program may be a failure resulting in a lost breeding season and the need to look for another mating next season.

With the mention of the foregoing breeding problems, it is easy to see why specialist breeders jealously guard their breeding stock; it has probably taken a great deal of effort to reach the stage of having together the correct breeding stock to consistently produce winners year after year. Many a true breeder is scoffed at by a visitor to their yards because the visitor often expects to see pens of top exhibition fowls and is disappointed to find that seemingly nondescript birds are in the breeding pen. The genuine breeder has probably found by much trial and error that the bird he has in the pen produces the offspring he seeks. Further to this, the breeder is most unlikely to part with such stock at any price, and it is futile to expect him to do so. It is better to listen carefully to what the breeder has to say about such stock and select similar birds from within your own stock. In such cases it pays to be a good listener.

5. Cockerel and Pullet Breeding

This is a strategy where specific birds are selected to produce the desired exhibition cockerels or pullets of a particular breed. Such pens do not produce only cockerels or pullets as some people think, rather 50% males and 50% females are produced as in any other hatching, but only the males or females are retained for exhibition purposes. The other 'unwanted' chickens are grown on to be selected later for suitable breeding stock when new future breeding pens are put together.

The reason for this occurring is that some breeds have become so specialised that many of the desired features in the exhibition stock antagonise one another when placed in the breeding pen, resulting in offspring that are neither like the exhibition male or female. In other words, where a breed requires separate cockerel and pullet breeding pens, two strains must be kept, and kept well separated in a breeding program.

Many breeders are not prepared to go to such trouble, preferring to concentrate on breeds that will produce exhibition males and females from one pen or what are called 'single pen breeds'. Where 'cockerel' breeding and 'pullet' breeding pens are required, this will be indicated in each of the breeds covered by this book.

Artificial Insemination

Artificial insemination (A.I.) involves the collection of semen (male reproductive fluid) from a rooster and using it to inseminate selected females, resulting in the female producing fertile eggs. In other words, it is man manipulated mating. A.I. requires a level of understanding of poultry physiology and the learning of certain techniques for it to be properly carried out. The process requires practice and the training of the participating fowls. It suits working with the assistance of a fellow poultry fancier as the various operations are easier performed with two sets of hands rather than struggling with fowls on your own, particularly large fowls.

Once mastered, A.I. is a useful breeding practice for the serious breeder for the following reasons.

1. With careful dilution of collected semen, up to 10 females can be inseminated with one semen collection. This means that a highly desirable male's influence can be spread over more females than allowing the rooster to naturally mate with females.

2. Breeds such as Pekins can still be shown whilst being involved in a breeding program.

3. Males in fat exhibition condition can be used for breeding. With natural mating, such a fowl could be a poor worker because of his condition.

4. Breeds of fowls that would find natural mating difficult can be successfully bred from using A.I., for example, in a program of 'breeding down' large fowls to produce the bantam version of a large fowl breed. Some breeders of large Indian Game find this technique useful as the breed has developed to such

an extent that the male finds it hard to effectively 'tread' the female.

The A.I. process can be divided into two parts.
1. Semen collection
2. Insemination of the female.

1. Semen Collection

Males have to be trained for effective semen collection and should be kept separated from females. For the actual collection of the semen, it is best to have two operators, one to hold the bird, the other to massage the male and collect the semen. The correct massaging technique varies from bird to bird and some practice is needed to find the most successful method. In general, the area of massage is the soft area of the abdomen just under the pelvic bones but others respond to massage further along the keel. Ejaculation is assisted by pushing the tail over the back. Long periods of massaging are counterproductive. If the bird does not respond after a reasonable period, put it back in the pen and try later. It is also important to quickly and quietly catch the bird or the massaging will prove ineffective. Birds that do not respond to training after a week or two should be rejected from the A.I. breeding program. Remember, only fully matured males should be used and not immature cockerels. With a properly trained bird, semen can be collected every second day, ensuring a good sized ejaculation suitable for dilution.

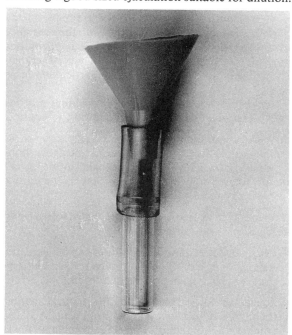

Equipment used to collect semen from males in an artificial insemination program. It consists of a small collecting funnel and a small glass tube connected by a piece of plastic tubing.

The equipment used to collect the semen is illustrated in an accompanying photograph. It consists of a small plastic collecting funnel around 5 cm in diameter across the top, a small glass tube and a piece of plastic tubing to connect the two. Alternatively if semen is not being diluted, it may be collected with the inseminator equipment. In this case, part of the collected ejaculation may be used to inseminate four or five hens provided sufficient care is taken when inseminating. However, if the semen is to be diluted, the first mentioned equipment has to be used.

If the semen is to be diluted, a Tyrode solution is used for this purpose. The normal dilution rate is about three times the volume of the collected ejaculant. That is, if 0.5 cc is collected, the final volume of the semen and diluent would be 1.5 cc. The Tyrode solution can be purchased in litre bottles and stored in the refrigerator. It can also be made up using the following formula

8.00 g	NaCl (sodium chloride)
1.0 g	NaH_2CO_3 (sodium bicarbonate)
1.0 g	fructose or glucose
0.2 g	$CaCl_2$ (anhydrous calcium chloride)
0.2 g	KCl (potassium chloride)
0.1 g	$MgCl_2$ (magnesium chloride)
0.05 g	NaH_2PO_4 (sodium dihydrogen phosphate)

These are mixed in one litre of distilled water.

Unfortunately, no long term method of storing fowl semen is available so collected semen must be used within a short period. The currently accepted maximum time for keeping is around five hours provided the semen is kept at 10–15°C, but this period is shortened in the warmer weather of summer when fertility tends to decline naturally. It is also known that storage life of semen varies from one male to another.

It appears that semen is best collected in the afternoon and the female(s) inseminated as soon after semen collection as possible.

2. Insemination of the Female

The equipment used to carry out this part of the A.I. operation is shown in the accompanying photograph. It consists of the glass end of an eye dropper and a small syringe which fits inside the glass eye dropper. A piece of rubber tube may be needed to fit over the needle mount of the syringe to facilitate a tight leakproof connection between the two. A small spring needs to be placed under the lip of the plunger handle, around the plunger shaft and onto the top of the body of the syringe. This is needed to apply pressure during the insemination of the female, or the semen is forced back out of the oviduct, making the insemination unsuccessful.

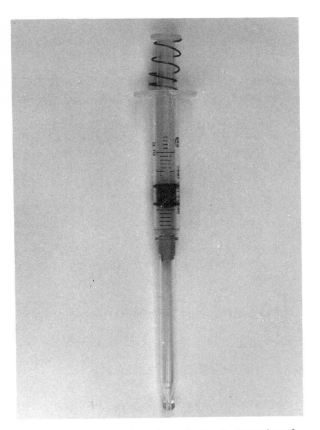

The equipment used to inseminate the female. It consists of the glass end of an eye dropper and a small syringe which fits inside the glass eye dropper.

An important step in the A.I. procedure is the eversion of the oviduct. This can occur if a female is simply picked up but others need a little encouraging by pushing their tails back over their backs. The everted oviduct can be held in the desired position with a little pressure being applied on the abdominal area surrounding the vent. As with the males, some practice is necessary to perfect the technique and train the fowl. Those where problems occur are best dropped from the A.I. program.

When the oviduct is in the everted position, the glass eye dropper is gently pushed into the oviduct, the pressure on the abdominal region needed to keep the oviduct everted is removed and steady pressure applied to the syringe as the inseminator is gently removed from the oviduct. This step also requires a little practice but is soon mastered.

0.1 cc of semen is necessary when indiluted semen is used but this then rises to 0.3 cc when standard diluted semen is used, that is 1 part pure semen and 2 parts diluent.

Inseminating equipment need not be sterilised after use and hot water will suffice when cleaning equipment. However, if soap or detergent is used in cleaning up, all traces of it must be removed as it adversely affects the sperm.

As can be seen, the A.I. procedure is really for the specialised breeder. It requires much 'hands on' practice to perfect the techniques and is made easier when working in conjunction with a fellow fancier on a specific breeding program. Difficulties aside, A.I. does open new opportunities for the serious breeder.

Ancona

Breed History

The Ancona breed takes its name from the Italian province from which the breed was imported into Great Britain in 1851. Evidence of the breed's existence can be traced back to the seventeenth century as it features in paintings of the time.

It was in Britain under the leadership of Bob Tunstall that the modern Ancona was developed and popularised. The utility properties of the breed were quickly recognised as well as its hardiness, feed economy and ability to forage well under free range conditions. Equally the Ancona adapted to intensive conditions and still remains to this day a popular backyard fowl.

An interesting early development was the fact that the Ancona won the only laying competition run in Britain for bantams. This utility bantam trend was continued for some time at some London shows around 60 years ago. The fad for utility bantams fortunately died out as most of the utility features of the breed were incorporated in the Standard and there was no need for a dual Standard.

The breed is also fortunate that it comes in one variety so that all the breeder's energies can be channelled into one direction.

On the Australian scene, W.L. (Bill) Marr has indicated that Anconas were first imported into this country on 17 May 1900. The Rosecomb variety was first imported pre World War I. The breed is fortunate to be served by an active breed club which started in 1948 and actively supports breed enthusiasts by often supplying special awards and bannerettes at important shows.

Whilst the breed is typically Mediterranean in nature, its flightiness can be overcome with good stockmanship.

Positive Features to Look For

Whilst at first glance the Ancona appears to have the same basic type as a Leghorn, there are features which give the Ancona its own distinctive type. The type we are looking for is broad at the shoulders, compact, with a medium length back tapering to the saddle. The front is rounded and the bird 'handles like a barrel' with a more fleshy feel than the Leghorn.

Specifically, the male we are looking for has a chest carried well forward, a back slightly sloping but straight, a neck which is long, arched and well hackled.

The tail should be carried at 45° to the back with plenty of furnishings. The tail feather should be wide and clearly marked with white tips. The wings should be carried well up against the body.

Turning our attention to the head, the face should be clean and bright with a full bright eye. The comb in the male should be of medium size, have a firm base and follow the line of the neck without touching it. Five to seven even serrations are called for by the British Poultry Standards, but it is the author's opinion

that birds with seven serrations have a tendency to be thin or pencilly and are less inclined to stand erect. The female's comb should fold gently to one side without obstructing the vision from the eye on that side of the head. The beak should be yellow with horn or a brown shading. The ear lobes are almond shaped, preferably kid white in colour. The legs need to be of medium length set well apart with the thighs almost hidden by the body feathers. Here too is a difference between the Ancona and Leghorn, the Ancona shows less thigh than the true Leghorn type. The leg colour is yellow mottled with black. The breeding of true leg colour is one of the more difficult challenges of the Ancona breed.

Likewise the attainment of the true standard colour requires skilful breeding. The ground colour is black with a beetle green sheen. Each feather should have a clear white 'V' tip to it. Anconas only come in one colour. The breed has also been bantamised.

Desired weights according to British Poultry Standards are as follows:

Large fowl
Cock 2.7 to 3.0 kg
Cockerel 2.5 kg
Hen 2.25 to 2.50 kg
Pullet 2.2 kg

Bantams
Male 570 to 680 g
Female 510 to 620 g

Negative Features to Avoid

1. The British Poultry Standards disqualifies roach back or any bad structural deformity.

2. Likewise, the British Poultry Standards rank the following defects accordingly:

(a) White in face
(b) Wings any other colour than black tipped with white.
Tail not tipped or black to the roots
Ground colour other than black with beetle green sheen
White or light under colour
Crooked toes
Squirrel tail
In knees (cow hocked)
(c) Bad comb

Group (a) attracts a 20 point deduction, Group (b) a 10 point deduction and Group (c) a 5 point deduction under the British Standard.

3. Other points that need to be looked for include:

(a) Combs which are overly large, thumb marked or showing side sprigs
(b) Tails that are narrow or whipped
(c) Face should be free of feathers
(d) Light or sunken eyes should be avoided
(e) Lobes that are yellow
(f) Feathers on legs or flat shins
(g) Black, grey, slaty or green coloured legs as these are often associated with poorly coloured feather tips. Quite often they are also poorly shaped tips.
(h) White in the flight feathers
(i) Too large a feather tip giving the fowl a 'gay' appearance
(j) Look carefully at the feather tips to see that they are not streaked with black giving a smudgy appearance
(k) Cull out birds showing purple, blue or bronze in the ground colour.

4. Bantams are prone to have low wing carriage and this should be rigorously culled.

Breeding Anconas

Anconas are a vigorous enough breed to stand line breeding for many years before an outcross is needed. Once the breed type has been fixed, it remains reasonably stable. The desired type has already been outlined but it is important that some points are again stated here when discussing breeding: a compact body, broad round front, broad shoulders, moderately long back and short thighs covered with body feather.

The Leghorn, when compared to the Ancona, has much more thigh, and a longer back line.

Once the type has been fixed, the problems of dealing with colour arise as 35 points are allocated to this breed feature. Such a large number of points would indicate that there is difficulty in attaining the colour as desired by the Standard and there is!

When breeding Anconas it is essential that birds be toe punched and leg banded to keep a track of them. It is important to know all birds' parentage and background if a breeder is to come to grips with this colour pattern. At the centre of the problem is the fact that birds can become lighter coloured or 'gayer' with each successive annual adult moult. Cockerels and pullets with the desired colouring may moult out lighter ('gayer') in the following moult. Dark specimens may similarly moult out to the desired colour the following year. Before attempting to tackle the changing colour situation, let's look at a standard mating. Cockerels for the breeding pen should show clean tipping, a good fine head, white tips on front and shoulders, saddle tipping and good white 'V' in

Single comb Ancona bantam female. Ancona Club of Australia Annual Show 1988. Owner: M. O'Connor.

Single comb Ancona large fowl male. Ancona Club of Australia Annual Show 1988. Owner: M. O'Connor.

Large Ancona female showing white markings required. Owner: A. Cheetham.

Large Ancona female. Camden Poultry Club Annual 1988. Owner: A. Cheetham.

Rosecomb Ancona female. Ancona Club of Australia Annual Show 1988. Owners: A. & S. Mills.

Single comb Ancona bantam male. Ancona Club of Australia Annual Show 1988. Owners: A. & S. Mills.

the tail. The pullets to be mated should have clean white tips *evenly* spread over the body, a soft folding comb, good white 'V' in the tail and strong yellow in the legs before laying. Both sexes need a rich beetle green in the black ground colour. Remember the pullets will be as dark as they ever will be.

A variation to this is to use cockerels with first year hens but you must know what these hens were like in their pullet year. Hence the need for good and accurate breeding records. All pedigrees must be known. The reverse of this approach, this is, using cock birds over pullets, can prove useful.

Another approach to selecting breeding pens is as follows. Some breeders prefer a 'balanced' mating where a darker bird is mated to a lighter one, but in doing this it is important to pay attention to *evenness* of colour distribution across the bird. Some breeders, when selecting the darker than exhibition male, also look for the darker spur colouring.

It is important that the weight be kept up in the females as quite often the well marked bird is also underweight. Breeding from such a female will cause a rapid loss of size.

Quite often when mating two exhibition coloured birds, the offspring have too much white in the flights and undercolour. Often they are overdone in the sickles of the resulting male progeny. This once again emphasises the need for good records, as once a good breeding pen is found it should be jealously guarded and not easily parted with. A proven breeding pen is invaluable to any fancier.

Some other plumage points that could prove of interest to the Ancona breeder are:

1. Look for even tipping across the back and breast
2. Select birds with a distinct 'V' tip and shun those with crescent shaped markings especially on the breast
3. Do not cull for colour or plumage until the adult plumage shows through in youngsters

4. Check the feather markings around the thighs as this is where the gay and dark ones show up
5. A dark pullet, provided the markings are correct, can moult out in her first hen year as a good specimen
6. Grisly flight and tail feathers lead to problems
7. Some breeders claim that early hatched birds are better furnished than their later hatched counterparts
8. Avoid females with red feathers
9. On the other hand, a little red in the wing bows and neck hackles of the male can produce the desired beetle green sheen in the ground colour especially in the pullets. However, an overuse of this can lead to red appearing more widely across the resulting cockerels and the possibility of developing a double mating situation
10. Hens that were good as pullets, colourwise, and resist 'gayness' in moulting should be highly prized as breeders.

This brings us finally to the problem of breeding the correct leg colouring in Anconas. This is a big challenge, that is, to obtain the black mottling on yellow legs. The additional problem with leg colour is that it may be good in the pullet or cockerel year but degenerate as the bird ages. Further frustrations occur when a breeder notices that often the yellow legged bird has too much white in the plumage or the tips are too large.

To add further frustration, it is often the bird with dusky coloured legs that has the sound flight feathers!

Breeding a top Ancona presents a challenge that would appeal to the keenest of poultry fanciers. If you are lucky enough to breed one, there are plenty of fanciers around to appreciate the beauty of such a bird, and the skill required to achieve it.

Andalusian

Breed History

The breed's name comes from the province of Spain, Andalusia, from where the early stock was imported into Britain for the subsequent breeding and selection to develop the Andalusian breed as we know it today. The original stock and its closely related neighbour, the Black Spanish, have been part of this Mediterranean region for many centuries.

The first of the Andalusian stock was imported into Britain as early as 1851 but were of a mixed appearance. They were first exhibited in Britain in 1853, and by 1888 there were 58 males and 55 females exhibited at London's Palace Show.

According to the British Poultry Standards, the blue was developed from crossing the black and white stock imported from Andalusia. This is not a true blue colour but more of a grey. Black Minorca blood was used to improve the lacing on the fowls.

This breed is classically quoted as an example of incomplete dominance in genetics. The breed appears in three different colour forms: black, white and blue. The blue fowls carry a black and a white gene; the black fowls carry two black genes; and the white fowls carry two white genes. When two of the desired blue fowls are mated, the offspring appears in the following proportions: 50% blue, 25% black and 25% white.

For this reason, the breed did not gain early popularity because it did not appear to breed true and there were large numbers of culls based on colour and poor lacing. However, the culls could have still served as good layers despite not being suitable for exhibition. No doubt they would have added some colour to early commercial or free range flocks!

It was Punnett and Bateson who first offered an explanation of the breeding pattern of the Andalusian in 1906. Had the early poultry breeders of this time grasped the explanation of the colour inheritance they would have mated the light (white) and dark (black) fowls to produce all blue offspring! However, the Andalusian breed never found the favour with commercial or utility poultry raisers that the other Mediterranean breeds did, such as the Leghorn, Ancona and Minorca.

The Andalusian breed does offer the fancier an attractive, alert, active and challenging breed to become involved with. They are non sitters and lay fine textured large white eggs. The eggs being large, well shaped, white and free of tinting are ideal for egg competitions that are sometimes run in conjunction with poultry exhibitions.

Little show preparation is needed, simply a head and leg washing prior to exhibiting.

Positive Features to Look For *(colour plates p.81)*

1. The Andalusian type is often confused with that of the Minorca but the form (shape) required is rather slim and reachy. The comb is smaller and finer than the Minorca. The ear lobes smaller and almond shaped. Plenty of thigh should show. The tail should be well furnished and carried well up.

2. As many feathers as possible should have the ground colour of slaty blue, some call it French grey, and an even border or lace of rich velvet black. To the breed's credit, the ground colour is sun and rain proof.

3. Whilst focussing attention on colour, the male's sickle feathers should be a dark even black, and hackles a rich lustrous black. The female should also have rich lustrous black neck hackle with broad lacing on the tips of the feathers at the base of the neck. The under colour in both sexes should tone with the surface colour.

4. The listed parts should show the following colours
Beak: dark slate or horn
Eyes: dark red or red brown

Lobes: white
Legs and feet: dark slate or black

The weights according to the British Poultry Standards are as follows:

Large fowls:
Male 3.2 to 3.6 kg
Female 2.3 to 2.7 kg

Bantams:
Male 680 to 790 g
Female 570 to 680 g

5. Much attention should be paid to the lacing. Equal lacing around each feather is the important aspect. A broad lace at the tip and narrowing down as it progresses up the feather should be avoided. In other words it should be even until it reaches the fluff. The exception to this is the neck hackle in the female at the base of the neck.

Negative Features to Avoid

1. The British Poultry Standards list the following as serious defects:

In the male:
White in the face
Red in the lobes
White feathers
Sooty ground colour
Red or yellow in hackles
Comb other than upright

In the female:
All of the above
An upright comb

2. The ground colour must be one clear shade of blue free from tickings and other additions of black.

3. Overdone combs and lobes which could be regarded more as what would be expected of a Minorca.

4. Be on the look out for any other evidence of back crossing with the Minorca such as
(a) sooty (black spots) in the ground colour
(b) sandy ground colour
(c) thick lobes
(d) indistinct lacing even though it is claimed that the Minorca was often used in the past to allegedly strengthen this particular feature
(e) white, red or yellow appearing anywhere in the plumage
(f) watch for light undercolour as this is an indicator of colour pigment deficiency and the ground colour will soon work out.

Breeding Hints

1. As with all breeds, the first priority in the breeding pen is type. The following assumes that type is in order and concentration is placed on colour and lacing which make up 50 points out of 100 points in the breed standard. Such a points allocation may indicate that type is of secondary importance but it should always be remembered that it is *type* that makes the breed.

2. There are two basic matings or pens that are used to produce exhibition Andalusians:

(a) The first is used strictly to breed for colour. That is two standard blue parents are mated. The results will be as follows:
50% will be blue more or less the same as the parents
25% will be too light, mostly whites splashed with black or dilute black in their plumage
25% will be either black or near black and carrying too much pigment.

(b) The second is a type based mating where the breed type of the parents is of utmost importance.
The birds to be mated will be a 'light' or 'white' bird with a 'dark' or 'black' bird. This pen will produce all 'blue' offspring and the normal process of shifting

through the offspring will be based on ground colour and lacing.

The top exhibition blues would go into the first mating with the knowledge that a good percentage of blues will result and the 'off colour' progeny could be used as a pool of birds from which the second pen could be selected. Of course, mating brother to sister is not encouraged as such a mating is too close, resulting in a lowering of size, fertility, vigour or stamina.

It can be seen that a success can be made of this breed with two carefully selected breeding pens.

Other 'type' based matings could be arranged as follows:

firstly, Black to Blue resulting in 50% Blue and 50% Black

and secondly, splashed White to Blue resulting in 50% splashed White and 50% Blue.

3. Now turning our attention to lacing. The inheritance of lacing is less well understood but some observations have been made which may assist in selection for breeding.

(a) Matings of birds with broad lacing that is not well defined produces birds with dark ground colour and lacing that runs into the web with a fuzzy outline and not the outline shape as required by the Standard.

(b) Balanced matings can be made using a wider laced bird to a finer laced bird with numbers of the progeny having the desired lacing width.

(c) Heavily laced, dark ground coloured birds when mated produce progeny that are too dark.

A top Andalusian, despite the breeding problems, is a real show stopper too few in numbers today.

Araucana

Breed History

Probably no other breed of fowl is shrouded in so much mystery and speculation as the Araucana from South America. For the poultry fancier, a breed that naturally lays blue shelled eggs would be an instant attention getter, and because of this the Araucana was quickly dubbed the 'Easter Egg Chicken'.

Its ancestors are only found in South America in Chile, Bolivia and Peru. Strong evidence suggests that the breed was present before the Spanish Conquistadors arrived there in the sixteenth century. The original bird is thought to have been rumpless and tufted. It also laid a range of egg colours from blue, azure, blue/green to a maroon dotted variation on a background of one of the preceding colours.

Today, in its country of origin, the most common fowl seen is a breed which has resulted from the interbreeding of Spanish breeds and the native fowl. It was this fowl, still possessing the blue egg laying capabilities, that was imported into both Great Britain and the USA.

The importations of birds and eggs took place from pre World War I to the 1920s. The imported fowls were similar in appearance to the Brown Leghorn, pea combed with light slate to near blue legs. The birds became the centre of various breeding programs including those which showed that the blue egg laying ability could be transferred to other breeds. Other breeds were also used to improve the Araucana's natural features, for example Houdans for improved tufting, Indian Game to strengthen the pea comb and the Blue d'Anvers Belgian to produce a lavender variety.

In the USA four types of Araucana were developed from the original imported stock:

1. Normal—blue egg layers, normal tails and normal ears (no tufts)
2. Tufted—blue egg layers, normal tails and having ear tufts
3. Tailless—blue egg layers, no tails and no ear tufts
4. Tufted Tailless—blue egg layers, ear tufts and no tails.

Fascinationg with the ear tufts intensified. The ear tuft was found to actually grow from the ear itself. A number of variations were produced:

1. backswept
2. frontswept
3. horizontal
4. tear drop or downwards.

Many new colours were developed mainly based on Old English Game colours.

This left the problem of standardisation of the breed in both Great Britain and USA. The Araucana is now accepted in the Tailed and Rumpless (Tailless) form. The standards in both countries permit large fowl and bantams. In 1976, the Americans accepted the following colours into the American Standard of Perfection:

1. Black
2. Black Red
3. Golden Duckwing
4. Silver Duckwing
5. White

In Britain, the favourite colour is Lavender but the following colours are also accepted:

1. Black
2. Blue
3. White
4. Black Red
5. Pile
6. Silver and Gold Duckwing
7. All Game colours excluding 'Off' colours.

In standardising the breed, the link between the pea comb and the blue egg laying ability was recognised. Both the British and American Standards call for a pea comb.

The breed is known to be reasonably productive with 180–190 eggs a year an acceptable expectation.

It has been claimed that blue egg laying fowls can be developed from the layers of brown shelled eggs as these eggs carry a combination of red, blue and yellow pigments. Breeding and selecting to eliminate the red

and yellow pigments will result in blue shelled eggs. How or why this took place naturally with the Araucana has remained a mystery. Whenever Araucana eggs are exhibited they still cause much comment and their colour is the central fascination of the breed.

Negative Features to Avoid

1. *Colour*
When breeding Lavender coloured birds, avoid straw colouring (brassiness) in the neck and saddle hackles, particularly in the males.

2. *Comb*
Cull out any comb other than pea comb.

3. *Cresting and tufts*
Avoid excessive cresting and tufts, especially mating the two together.

4. *Wings*
Wings should not be carried too low.

5. *Backs*
Avoid round backed birds.

6. *Tails*
Cull out birds with tails being carried too low or too high. A 45° angle is desired.

Breeding Hints

The two main areas of interest with Araucana breeding are:
• blue egg laying
• tufts and lethal gene combinations.

Blue Egg Laying
As mentioned previously, the pea comb and the blue egg laying ability are linked. Single, rose and walnut combs do not breed true to blue egg laying. There is no connection between ear tufts and blue egg laying ability. Likewise, taillessness has no connection with blue egg laying ability.

However, if a blue egg layer is crossed with a white egg layer, the female offspring lay blue shelled eggs. If a blue egg layer is crossed with a brown egg layer, the resulting females could produce eggs of the following colours: blue, near blue and blue green.

Ear Tufts and Lethal Gene Combinations
Crossing tufted birds produces lethal gene combinations. Homozygous for tufts produces 100% lethal and 20–25% lethals can result from heterozygous tuft matings. These lethals are expressed as 'dead in the shell' at hatching. When these dead embryos are investigated they are found to have 'hole in the head' deformities such as large holes in the throat, small holes below the ear and elongated ear openings.

Another investigation found amongst the 'dead in the shell' the following range of deformities:

1. no apparent deformities
2. small hole through the side of the head just below the tuft
3. eyeless
4. unhinged beak
5. rear extensions of beak protruding through holes in the throat or side of the head
6. tufts growing on the ends of the beak.

It was noted that the deformities were only associated with tufted embryos. Deformities were most pronounced when heavily tufted males and heavily tufted females were mated.

Current scientific thought suggests that tufting is controlled by an autosomal dominant gene which causes prenatal death to all homozygous individuals and 20% of all heterozygous individuals.

Australian Game

Breed History

The Australian Game fowl appears to have common beginnings with the Australian Pit fowl, but through differences in breed infusion and selection the Australian Game was developed along different lines. Both breeds are based on a mixture of Old English and Asian blood but the use of the Malay seems to have played a more important part in the Australian Game's development, producing in the early days long legged fighting fowls called Colonials.

With the banning of cockfighting, the Colonials were further developed by the farmers of the Windsor/ Richmond (Hawkesbury district) and West Maitland (Hunter district). The ideal of pit (fighting) fowls was replaced by a desire for table (utilitarian) and exhibition fowls. They were then given the name Colonial Game. Isaac Hopkins of Windsor is accredited as the first to breed the Colonial Game for exhibition, and his fowls were exhibited at the first show held by the Royal Agricultural Society of N.S.W. in Parramatta Park. Their appearance in the show pen coincides with the fad in the poultry world at the time for long legged, tall stationed (reachy) breeds such as the Modern Langshan, Modern Old English and Modern Cochin.

Its period of greatest popularity as an exhibition fowl was from 1880–1915. Harry Maude, the legendary poultry expert, recalled as a young fellow seeing 300–400 entries being penned at the Hawkesbury Show in the early 1900s. This figure becomes more impressive when one considers that some of these entries were for a pen of 12 table fowls! This show was regarded as the Mecca of the breed with competition becoming so intense that feather plucking and faking became commonplace. Judges were subsequently instructed to pass any birds showing evidence of faking. The impact of the faking has left its mark on the breed Standard which specifically spells out that any bird showing signs of being faked is to be disqualified. In the heyday of the Hawkesbury Show, birds changed hands for large sums of money. But, after World War I, numbers declined, demand decreased and prices fell sharply.

At the peak of the breed's popularity, many cups and trophies were offered at leading shows. Perhaps the most prestigious was the Greville Challenge Cup. In 1897 this was won by a 5.45 kg Black Red cock exhibited by a Mr W. Webb. This fowl also won the trophy for the heaviest fowl exhibited at the show. At the same show, Australian Game fowls won first prize in the 'table poultry for export class'. Such was the popularity of the breed that in 1889 John Fairfax and Sons commissioned the well known feather artist of the day Neville Cayley to produce black and white illustrations for publication in the *Sydney Mail* newspaper, of Mr Webb's winning Black Red cock bird and pullet, to show the ideal of the breed. The fowls were promoted as being 'good layers of well flavoured eggs, the best of sitters and mothers'—contrary to the view held now—'hardy, excellent foragers on free range and excellent table birds'. In other words, the ideal farmer's fowl.

The first breed Standard was drawn up by a Mr Silcock and adopted in 1896. An 1892 winning hen owned by Mr T. Campbell of 'The Oaks' was used to illustrate and describe the colour standard for the Black Red female after rejection of the female colour pattern featured in Lewis Wright's classic plates.

The Australian Game came in many colours with the following being documented:

1. Black Red
2. Golden and Silver Duckwing
3. Pile
4. Black
5. White
6. Henfeathers

Some Cuckoo and Golden Cuckoo have also been reported.

Today, the breed is a rarity in the hands of very few breeders and is in danger of being lost forever unless more supporters can be found. A bantam form also exists in very small numbers.

Positive Features to Look For *(colour plates p.82)*

The accepted Standard for Australian Game as published in the *Poultry* newspaper 10 October 1959 is as follows.

General Characteristics of the Cock

1. Head: Long and strong
2. Beak: Strong, slightly curved, stout at base
3. Comb: Pea or triple (if undubbed—surgically trimmed). In no case to go further back than a point directly above the back of the eye
4. Face, ear lobes and throat: Fairly smooth
5. Eyebrows: Slightly overhanging
6. Neck: Long and slightly arched
7. Hackle: Short and close fitting
8. Saddle hackle: Short
9. Body: Short, stout, widest at shoulders and tapering to tail
10. Back: Flat
11. Shoulders: Broad, high and square
12. Breast: Hard, broad and full
13. Wings: Medium length, strong, well clipped under the saddle hackles
14. Thighs: Prominent, set wide apart, long, stout and muscular
15. Shanks: Evenly scaled and slightly rounded
16. Spurs: Set low down on the shank and inclined downwards
17. Feet: Flat on the ground
18. Toes: Strong and well spread, the hind toe well extended and flat on the ground
19. Tail: Medium in length, slightly drooping in carriage, and carried moderately full; sickle feathers fairly abundant and slightly curved
20. Plumage: Sound, glossy and hard
21. Size: Large, minimum weight of adults, 4.09 kg
22. General appearance: Upright, active and reachy.

General Characteristics of the Hen

The hen should resemble the cock in all points, making allowance for difference in sex.
Minimum weight of adults 3.18 kg.

Standard of Points for Judging

A bird perfect in shape, size, colour, comb and condition to count 100 points. Relative importance of defects and order in which they are to be taken into consideration in judging are:

Points to be deducted

1.	Want of symmetry and strength	15
2.	Want of size	10
3.	Want of condition	10
4.	Want of hardiness in handling	10
5.	Bad or faulty plumage	10
6.	Less too thin (5) flat shinned (5)	10
7.	Tail too heavy (5) badly carried (5)	10
8.	Too much hackle	10
9.	Bad head	5
10.	Bad feet and legs	5
11.	Bad coloured eyes	5
		100

Colour Descriptions according to the late A.J. Compton

1. Black Red

In both sexes
Beak: Dark horn colour
Face, comb, earlobes, wattles: Deep rich red
Eyes: Pearl, yellow or daw (pearl coloured)
Shanks and feet: Willow or olive green

The cock
Head and hackle: Rich deep red
Back, shoulder coverts and wing bows: Deep dark red
Wing bars: Steel blue
Breast and wing butts: Greenish black
Underparts: Black
Wing primaries: Black with edging of bay to outer web on the lower feathers.
Wing secondaries: Outer web deep rich bay with inner web black with a blue–black tip on the end of each secondary feather
Saddle: Deep red
Tail: Glossy green-black

The hen
Head: Brown
Neck: Yellow
Neck hackle: Striped with black in the centre of the middle and lower feathers beginning at the back of the head
Throat and breast: Dark salmon colour, running to a lighter shade on thighs and vent. Each feather showing shafts of a lighter shade
Body: Dark brownish drab or deep solid partridge brown, evenly and minutely pencilled with markings of black. This marking should have an irregular or wavy appearance, and on no account should run into distinct bars
Tail: Black. The side coverts and the two top outer feathers should be the same shade and markings as the body colour.

2. Duckwing
In both sexes
Beak, face, comb, earlobes, wattles, eyes, shanks and feet, same as in Black Reds.

The cock
Breast, tail, wing bars and underparts, same as in Black Reds
Neck hackle and saddle hackle: Clear creamy white
Back, shoulder coverts and wing bows: Deep rich gold
Primary wing feathers: Black with edging of white on lower feathers
Secondary wing feathers: Clear white on outer web, black on the inner, the tips of each secondary feather being marked with steel blue

The hen
Head and neck hackle: White with clear and distinct black stripes on each side of a white shaft down the centre
Back and wings: Grey or ash colour, evenly and minutely pencilled with minute black wavy markings. This marking should have an irregular or wavy appearance and on no account should approach distinct barring
Breast: A lighter shade of salmon than the Black Red hen
Thighs and underparts: Also being of a greyer shade in the black
Tail, side coverts and the two top outer feathers: Same as the body colour and minutely pencilled with black.

3. Pile
In both sexes
Beak, shanks and feet: Yellow or willow (the former preferred as it matches the light plumage of the Pile)

Face, comb, ear lobes, wattles and eyes: Same as for the Black Red

The cock
Clear, rice white, where the Black Red cock is black, and red where he is red.

The hen
Head and neck hackle: Golden, with a white stripe down the centre of each feather
Back, wings and tail: Clear white
Breast and throat: Dark salmon, running off to a lighter shade on shafts.

4. Black
In both sexes
Beak, eyes, shanks and feet: Black
Face, comb, ear lobes and wattles: Rich deep purplish red
General plumage: Metallic glossy greenish-black throughout.

5. White
In both sexes
Beak, shanks and feet: Yellow
Eyes: Pearl, yellow or daw (pearl coloured)
Comb, ear lobes and wattles: Rich bright red
General plumage: Pure spotless rice–white throughout.

6. Henfeathered
The cock
Feathered as hen, and in both sexes colour and markings exactly the same as the hens of the various varieties such as Black Red, Duckwing and Pile.

Negative Features to Avoid

These can be divided into physical and colour faults to be avoided. At this point the physical faults will be dealt with and the colour faults as they occur in the section dealing with hints on breeding.

The breed Standard lists a number of faults which are regarded as disqualifications and they include:

1. Roach back
2. Crooked breastbone
3. Duck feet
4. Wry tail
5. Any other deformities or disease
6. Comb other than triple or pea on hen
7. Feathers on shanks or feet
8. Any fraudulent dyeing, dressing or trimming beyond the dubbing and dressing of the heads of cocks.

Other faults that must be guarded against are

1. Flat shins
2. Goose wings
3. Twisted or curved toes
4. In knees (cow hocked)
5. Closely whipped tails as seen in Modern British Game.

Breeding Hints

As with all breeds, the Australian Game must be physically sound to be placed in the breeding pen. Because of the leggyness of the breed, they do not stand being confined to small breeding pens and do best where they can range or be given a much larger than usual deep litter pen in which they can adequately exercise.

In the days prior to modern scientifically developed foods, many Australian Game breeders had their own formulae for feeding the fowls, particularly the youngsters. A sensible feed program of chick or grower crumbles, green feed and some grain at the appropriate time would more than adequately meet the needs of the young growing stock. An important requisite is space. This is needed for exercise and to minimise the social squabbles that invariably occur with all Game fowls. The size must be kept up. Make sure that it is not just fat but muscle and frame.

1. Black Reds
Before specifically turning our attention to the make up of the breeding pen, two faults that must be guarded against: yellow legs and white in the face. Black Reds showing these faults must be culled.

Both top quality males and females can be bred from the one pen, provided they show good type, size and are well coloured.

Points to look for are:

Male: even colour across the top from head to tail. The breast must be free of blotches, spangles, red lacing or rust colour.

Female: a dark partridge colour free of rust, foxy or rosy colouring on the back or wings. Make sure there is plenty of shaft (light quill) in the body feather.

A fault to avoid is the wheaten colour which has been introduced by the re-infusion of Red Malay males or Wheaten Malay females into breeding stock at some earlier time.

2. Duckwings
These are reported to be more difficult to breed to true colour than Black Reds. Unless the males are kept out

of the sun and rain after moulting, the neck and saddle hackle feathers turn brassy. It is also important to remember that lacing and flecks appear in males as they age.

Some further points to bare in mind are

(a) Avoid striping in the neck hackle
(b) Avoid any chocolate colouring on the outer edge of wing feathers
(c) Choose steel grey coloured females as those with smoky or cloudy body colour generally have coarse markings or blotchiness in the flight feathers.
(d) Guard against foxiness or rust in wings or body
(e) Despite best endeavours, the female colour generally lightens off towards the thighs.
(f) Females with very silvery neck hackles breed males with little or no stripe or streakiness in the neck hackle.

Matings used are as follows:

(a) A good Black Red male to a Duckwing female
(b) A well coloured Black Red female to a Silver or Golden Duckwing male.
(c) Well coloured Golden Duckwing cockerels can be produced from mating a Golden Duckwing with a rusty winged Duckwing female
(d) Good coloured females with a tendency to run light can be produced from mating a good Duckwing female to a Silver Duckwing male.

3. Piles
Double mating is required here to produce top quality males and females. Double mating meaning one pen is used to produce desired females and a differently mated pen to produce the desired males.

Clean roomy, deep litter conditions with access to a clean grass run are the desired conditions to raise this variety. Ideally, to maintain their colour in immaculate condition, Piles should be kept out of the weather.

A.J. Compton claimed the way to produce top quality piles was as follows:

(a) *Cockerel breeding pens*
Put together to produce the desired coloured cockerels.

Pen one: Use Black Red blood every third year to keep the colour of the cockerels up. In doing so, the leg colour is also maintained. A bright coloured Black Red male with clear hackles is mated to Pile hens with light salmon breasts and neck hackles very lightly edged with pale gold. The rest of the female body and tail should be as clear as possible.

Pen two: Use a good coloured Pile male with hens heavily marked with gold in the neck hackle, pale in the breast and wings, and backs well mottled with rust or brown feathers.

2(b) *Pullet breeding pens*

In this case use a male that is not as brightly coloured, more a chestnut red in the hackles and back. Look also for a few breast feathers that are lightly laced with gold but make sure that the wings are sound. To this male should be mated a female with a sound deep salmon breast, clearly marked gold and white hackles and a pure clear white body colour. The next season, put the best pullet(s) back to the father and the best hen(s) back to the best cockerel.

If space only allows one pen, the following mating will produce some exhibition fowls of both sexes. Use a golden orange coloured cockerel with sound wings to a foxy winged female plus a clear winged, sound breasted female.

4. Blacks

This variety no longer exists but Compton claimed that they were bred by using cockerel and pullet breeding pens. The cockerel breeding pen consisted of mating two exhibition coloured fowls and the pullet breeding pen used an exhibition female with a male showing coppery red in the neck and saddle hackles.

5. Whites

These were never very popular but are being mentioned for historical value. Maude claimed that they were produced as sports from Pile and Duckwing matings. Compton gave the breeding tip of making sure that the breeding pen was made up of fowls which were of the purest white free of any other colours.

6. Henfeathers

Little is known on how these are bred. They were known to exist in Partridge and Duckwing colour.

Australian Game Bantams

Australian Game bantams do exist with Reg Tutty of Greensborough in Victoria being responsible for their development and subsequent standardisation in Victoria. Reg Tutty claims in his recent book *My Lifetime in the Game Fancy* that the bantams are as good in type as the large fowl. They are to be found in Black Red, Pile, Duckwing and Golden Crele, so the remarks already made on colour would also apply to the bantams.

Australian Pit Game

Breed History

The Australian Pit Game was developed by British officers stationed in early colonial Australia and in particular around Parramatta. These officers would have been influenced by cockfighting in England and what was seen on their tours of duty, principally in India. As with most early Australian settlers, the soldiers wished to have and do the things they were familiar with in the home country, and so cockfighting found its way to Australia. It could be speculated that the early officers, having seen both the genuine Old English Game fowls of England and the fighting fowls of the Indian subcontinent, set about breeding fowls that combined the desired fighting features of both types.

Ralph Berman, noted game fowl expert, states that the Australian developed pit (fighting) cock of the time was a combination of Oriental and Old English blood lines. Harry Maude, the legendary poultry expert of the first half of this century, states that the Australian Pit Game fowl was made up of Modern British, Malay and Old English with a later infusion of the Asiatic breeds the Aseel from India and the Sumatras from the Philippines.

What the early cockers were trying to do with this crossbreeding was to breed a short heeled (short steel spur) bird with the speed of the Old English and the rugged strength plus staying power of the Oriental. With such crossbreeding came diversity of type depending on what mixture of breeds was used.

The pit or fighting fowls were taken wherever the early soldiers were stationed in eastern Australia. Their fighting ability soon left its mark. Many settlers took on the breed which started to express itself in different types linked to the area in which they were bred. Three well known types based on area of development were the Queensland, New England and Central Coast (N.S.W.) types. This later posed problems when attempts were made to standardise the breed for exhibition. It was thanks largely to the work of Marshall, Compton and Pearson that an Australian Standard was finally agreed upon in all States. This Standard was included in the classic work of Marshall *The King of Fowls* and remains the standard by which Pits are judged today.

Positive Features to Look For

1. Weight: Male to 4.54 kg
   Hen 1.82 to 3.18 kg
2. Head: Powerful, medium in length
   Skin of face and throat smooth and fine in texture, loose and flexible at throat but not flabby
3. Comb: Single comb/variety: small, erect, firm, straight and evenly serrated
   Pea or triple comb variety: small, the central division slightly higher and longer
   Dubbing (surgical removal of wattles and ear lobes, and shaping of comb) is allowed in male birds
4. Ear lobes and wattles: Smooth and close fitting
5. Beak: Stout and strong at base, well curved and pointed, the upper and lower mandibles locking together like a vice when closed
6. Eyes: Large, bold, fiery and fearless
7. Neck: Fair length, strong boned and slightly arched, well developed at junction with body
8. Body: Heart shaped, back fairly short and flat, broad at shoulders, with deltoid muscles well developed but maintaining the shoulder width across the back from thigh to thigh tapering to a fine stern at the set of the tail. The body on top shaped like a smoothing iron with full and well rounded sides
9. Shoulders: Well braced, high and prominent without any tendency to hollowness between

10. Breast: Broad, prominent and full with the pectoral muscles strongly developed, curving under sharply to show clearly defined junctions of the thighs

11. Wings: Fairly long and powerful
Wingbows well rounded
Wingbutts stout and prominent
Secondary and primary wing feathers with strong quills, well webbed and folded tightly. The secondary wing feathers covering the primary wing feathers when wings are closed

12. Belly: Compact and tight
Free as possible from fluff feathers, with a clean run behind from back of thighs to the vent

13. Tail: Medium length
Carriage moderately elevated but 'governed by the character of the bird'

14. Legs: Medium in length to 'suit character of the bird'
So placed as to ensure movements of force and activity

15. Thighs: Medium in length
Set fairly well apart
Well developed and muscular
Curved at junction with the body and tapering off to clean bone and sinew immediately above the hock joint

16. Hocks: Slightly bent but not to give a crouched appearance

17. Shanks and feet: Clean, fine and strong boned of medium length
Round in front and flat sides with wiry tendons showing
Even and close fitting scales

18. Toes: Four in number, well spread apart
Medium length with powerful toenails
Hind toe carried well back from the leg and nearly flat on the ground

19. Spurs: Hard, set low down on leg
Cutting permitted

20. Plumage and furnishings: Hackles, true tail, sickle and side hangers to be of moderate length and fullness 'according to the character of the bird'.
Each section to harmonise with the other

The whole plumage throughout to be hard (close fitting), sound, resilient, smooth and lustrous
Muffs, tassels and henfeathers eligible for competition

21. Colour: Immaterial

22. General shape and carriage: Proud, defiant and aggressive looking
Movements quick and graceful, ready for any emergency
Vigorous, alert and agile

23. Handling: Body well balanced
Hard, firm, yet somewhat light fleshed, corky, mellow and warm
Strong contractions of wings and thighs to the body when in hand.

Female
The hen of each variety should resemble the male in each essential point making allowance for sexual differences.

One phrase which appears several times is 'character of the fowl' and Marshall explained the use of this phrase by reminding people of the fact that the Pit fowl had a mixed background. Some had more Old English Game in their make up where others tended more towards the Asian stock. To cover this situation, Marshall used the phrase 'character of the fowl' when birds were being assessed by the Standard.

To summarise in the words of the late Harry Maude, what we should be looking for in order of importance is:

1. The greatest degree of 'gameness' or courage, combined with the activity and strength, or physical fitness to uphold same in battle fighting against an opponent similarly equipped.

2. Formation of shape should be such that every ounce of weight lies where it is calculated to ensure the greatest degree of activity and strength, which would mean that the body should be rather short for its width, wide at the shoulders, and wide towards the

Australian Pit Game bantam male. Owner: J. Clark. Bantam Club of N.S.W. Annual Show 1988.

Australian Pit Game female. Fairfield Poultry Club Autumn Show 1988. Owner: J. Gardner.

Australian Pit Game male. Fairfield Poultry Club Autumn Show 1988. Owner: J. Gardner.

Large Australian Pit Game male. Camden Poultry Club Annual 1988. Owner: R. Thelan.

stern, where it is clean cut, full in front and in girth, and tapering away; handling light and corky but muscular.

3. The general make-up should be active looking and strong but not too puggy or cobby—in other words built more like a gladiator or all round athlete than a heavy, cumbersome weight-lifter. Legs should be of medium length, medium in bone substance, hard, sinewy, clean and smooth of surface, and bent at the hocks; toes, long and tapering, the back or prop toe of good length, low set, and almost in a line with the middle toe—an important essential.

Scale of points

Defects	Deduct up to
Head 4; beak 5; eyes 6	15
Neck 6; back 9	15
Breast and body	15
Wings	8
Thighs 5; shanks 6; spurs 3; feet 9	23
Carriage	6
Plumage	8
Handling	10
A perfect bird to count	100

Negative Features to Avoid

The Standard lists the following serious defects for which birds may be disqualified.
 1. Thin thighs or neck
 2. Flat sidedness
 3. Deep keeled, pointed or badly indented breastbone
 4. Thick insteps or toes, flat shins
 5. Duck heels
 6. Straight or storky legs, bow legs, 'in-knees'
 7. Soft flesh, soft on handling
 8. Faulty or fluffy plumage
 9. Clumsy carriage or action
10. Wry tail
11. Slip wing
12. White in lobe
13. Roach back.

Some other points that need to be weeded out are

 1. Baggy bellies
 2. Drooped or squirrel tails
 3. Stilty or stubby legs
 4. Hollow backs.

Breeding Hints

1. Do not have the same faults on both sides of the mating. Try to balance weak points with strength in that point in the breeding partner. Much emphasis should be placed on the female type— she has to be right.

2. Line breed the fowls to ensure the desired features remain in the offspring. Outcrossing should be attempted only when the line has run into a dead end.

3. Cull hard for fitness. Eliminate those that are unhealthy, unthrifty, those with bad feet, roach back, wry tail, crooked breastbone and those of non game type.

4. Do not be tempted to introduce Indian Game blood into the line in an attempt to widen the shoulders as a line of 'waddlers' could result instead of true game fowls.

5. The Pit chickens are very aggressive even from an early age so much space and close supervision is needed to rear them.

The breed also exists in bantam form so the foregoing remarks would equally apply to them, bearing in mind the smaller size of the bantam fowl.

Autosexing Breeds

Breed History

These breeds were developed after the initial work by Punnett and Pease at Cambridge University when in 1929 they produced the first of the autosexing breeds, the Cambar, from crossing the Golden Campine and the Barred Plymouth Rock. The Cambar was a rather nondescript breed which did not prove popular. However, the ideas behind the breed's development were not lost and when more commercially useful breeds were incorporated into subsequent breeding programs, a wave of sex linked crosses and new autosexing breeds swept across the poultry scene, particularly the commercial sector.

It was noted when these breeds were developed that chickens could be sexed at day old by their colour and there was no need for the vent sexing of chickens. The work of Punnett and Pease proved very useful during World War II as chickens could be sexed at day old and male layer chickens eliminated, thus saving scarce fowl food. Coupled with this, it was noted that the resulting chickens possessed much hybrid vigour, particulary in regard to growth rate. It could be speculated that this was the turning point in the poultry industry when separate lines of specialist laying and meat producing chickens were developed, marking the imminent end of the commercial poultry industry relying solely on pure breeds for their production birds.

Before venturing further, a clear distinction has to be made between the terms 'sex linked' and 'autosexing'. Here sex linked refers to the mating of two different breeds in order to recognise the sex of the progeny at hatching. Such crosses involve

1. 'Gold' and 'Silver' parents which produce chicks whose sex can be recognised by down colour at hatching
2. 'Barred' and 'Unbarred' parents where the male chick has a white head spot and the female does not
3. Parents that produce chicks with different coloured shanks, generally the male being yellow and the female being slate or dark coloured

4. Parents producing fast feathering (wing) females and slow feathering males.

On the other hand, 'autosexing' means mating of fowls of the same breed with the sex of the chickens recognisable at hatching. Examples of such breeds including the Legbar to follow later.

Examples of sex linked crosses, that is two different breeds producing chickens whose sex can be recognised at hatching include the following:

1. Rhode Island Red male × Light Sussex female. The resulting pullets are a warm buff shade and the cockerels are white or light yellow. It is important to remember that this cross must be made this way around or no difference is noticed in the resulting chickens.

2. Brown Leghorn male × Light Sussex female. The pullets have a light brown head and back with heavy chocolate stripes down it. Often two small thin stripes of similar colour run down either side of the main stripe(s). The cockerels are black and white.

3. Buff Leghorn male × Light Sussex female. Pullets are buff coloured with occasional slight black markings whilst the cockerels are silvery white also with the occasional black markings. The ground colours, however, are readily distinguishable.

These three examples are those of a 'gold' and 'silver' cross.

4. Black Leghorn male × Barred Rock female. The pullets have a pure black head and the cockerels a distinguishing white spot on the top of the head. An Australorp or Black Langshan male can also be used in this cross with similar results.

5. Barnevelder male × Columbian Wyandotte female produces similar results to the Brown Leghorn male and Light Sussex female cross.

6. Buff Orpington male × Light Sussex or similar silver female follows the 'gold' and 'silver' cross.

7. Brown Leghorn male × Barred Rock female.
The cockerels are black on the top of the body with a white head spot. The beaks, shanks, and toes are yellow. The pullets are all black on top with beaks, shanks and toes black or very dark.

8. Golden Legbar male × Light Sussex female cross produces golden coloured pullets and silver coloured cockerels.

9. Golden Legbar male × Rhode Island Red female produces willow legged pullets and yellow legged cockerels at hatching time.

Both crosses 8 and 9 show double sex linkage as the pullets have fast feathering (wings) whereas the males show little or no flight feather development just after hatching.

10. Golden Legbar male × Brown Leghorn female cross produces chicks separated by shank colour, the pullets being willow coloured and the cockerels yellow.

11. Rhode Island Red male × White Wyandotte female produces pullets which are brown with a darker stripe running down them. The cockerels are silver with black markings.

Other sex linked crosses attempted include Black Leghorn male (or any true Black male) crossed to a Legbar female. Similarly the Brown Leghorn male and White Wyandotte cross shows sex linkage. In this cross, any autosexing breed male can be used in place of the Brown Leghorn male and sex linkage will occur.

Following the work of Punnett and Pease a craze swept the poultry world for developing autosexing breeds most of which ended up having little commercial significance. The following is a list of some of them:

1. Gold and Silver Cambar
2. Gold and Silver Legbar
3. Gold and Silver Welbar (Barred Welsummer)
4. Gold and Silver Dorbar (Barred Dorking)
5. Brockbar (Barred Buff Rock)
6. Buffbar (Barred Buff Orpington)
7. Anconbar (Barred Ancona)
8. Gold and Silver Oklabar (meat type chicken)
9. Hampbar (Barred New Hampshire Red)
10. Gold and Silver Brussbar (Barred Brown Sussex)
11. Gold and Silver Wybar (Barred Laced Wyandotte)
12. Rhodebar, also known as a Redbar (Barred Rhode Island Red)
13. Arabars—developed from Araucanas
14. Australbar also known as Silver Lea (Barred Australorp)

Interestingly, it is claimed that the autosexing breeds are sex linked when crossed between themselves, with the pullets having characteristic striping and the cockerels a large head spot.

The Legbars were the most widely commercially accepted autosexing breed and some still exist in the hands of a few enthusiasts today.

Legbar

Because the Legbar performed to commercial expectations it did gain some acceptance. It was bred as Golden Legbars, Silver Legbars and a little known variety called the Cream Legbar, which was reported to lay blue eggs.

Positive Features to Look For *(colour plates p.82)*

Basically, the Legbar is a 'barred brown Leghorn', so the important thing to look for is the Leghorn body type; the Plymouth Rock brick shape should be avoided. The standard for the breed published in the late 1950s in the *Poultry* newspaper is as follows:

General Characteristics
Male
1. Carriage: Very sprightly and alert
2. Body: Wedge shaped, wide at shoulders and narrowing slightly at root of tail
3. Back: Long, flat and sloping slightly to the tail
4. Breast: Prominent
5. Breastbone: Straight
6. Wings: Large, carried tightly and well tucked up
7. Tail: Moderately full, at an angle of 45° from the line of the back
8. Head: Fine
9. Beak: Stout, point clear of the front of the comb
10. Eyes: Prominent
11. Comb: Single, perfectly straight and erect, large, deeply and evenly serrated (5 to 7 spikes), extending well beyond the back of the head and following, without touching, the line of the head
12. Face: Smooth
13. Ear lobes: Well developed, pendent, smooth, equally matched in size and shape
14. Wattles: Long and thin
15. Neck: Long and profusely covered with feathers
16. Legs: Moderately long
17. Shanks: Strong, round and free of feathers
18. Toes: Four, long, straight and well spread
19. Plumage: Silky texture, free from coarse or excessive feather
20. Handling: Firm, with abundance of muscle.

Female
Similar to the male, allowing for the natural sexual differences, except that the comb may be erect or falling gracefully over either side of the face without obstructing the eyesight, and the tail should be carried closely and not at such a high angle.

Colour

Golden Variety
Male
1. Neck hackle: Pale straw, sparsley barred with gold-brown
2. Wing coverts (or wing bar): Dark grey barred
3. Primaries and secondaries: Dark grey barred, intermixed with white, upper web of secondaries also intermixed with chestnut

4. Saddle hackle: Pale straw barred with bright gold-brown, as far as possible without black
5. Breast and underparts: Dark grey barred
6. Sickles: Paler
7. Tail coverts: Grey barred.

Female
1. Hackle: Pale gold, marked with black bars
2. Breast: Salmon, clearly defined
3. Body: Dark smoky or slaty grey-brown with indistinct broad soft barrings. The individual feather showing paler shaft and slightly paler edging
4. Wings: Dark grey-brown
5. Tail: Dark grey-black with slight indication of lighter broad bars.

Silver Variety
Male
1. Neck hackle: Silver, sparsley barred with dark grey but tips of feathers fade off to pure silver
2. Saddle hackle: Silver, barred with dark grey, the feathers tipped with silver
3. Back and shoulder coverts: Silver, with dark grey barring, the feathers tipped with silver
4. Wing bow: Dark grey with silver-grey barring
5. Primaries: Dark grey with some white permissible
6. Secondaries: Dark grey with tips of upper web white
7. Breast: Evenly barred dark grey and silvery grey with well defined outline
8. Tail and tail coverts: Evenly barred dark grey and silver-grey, sickles being paler.

Female
1. Head and neck hackles: Silver, with black striping, softly barred grey
2. Breast: Salmon, clearly defined
3. Body: Silver-grey, with indistinct broad soft barring, individual feathers showing lighter shaft and edging
4. Wings: Silver-grey
5. Primaries: Silver-grey
6. Secondaries: Silver-grey, upper web a lighter grey mottled
7. Tail: Silver-grey with indistinct soft barring.

In both sexes
1. Beak: Yellow or horn
2. Eyes: Orange or red, pupils clearly defined
3. Comb, face, wattles: Bright red
4. Ear lobes: Pure opaque white (resembling white kid) or cream, the former preferred. Slight pink

markings and pink edging not unduly to handicap an otherwise good bird for *utility purposes*.

5. Legs and feet: Yellow or orange or light willow in the female.

Standard weights

Cock:	3.18 to 3.41 kg
Cockerel:	2.73 to 2.95 kg
Hen:	2.27 to 2.73 kg
Pullet:	2.05 to 2.27 kg

Negative Features to Avoid

The Standard lists the following defects for which a bird may be passed:

1. Side sprigs on comb
2. Eye pupil other than round and clearly defined
3. Crooked breast
4. Wry tail
5. Any bodily deformity

Serious faults which should be culled out include:

1. Male's comb twisted or falling over. Overly large combs. Thumb marks in male's comb
2. Ear lobes wholly red. Any showing folds
3. Any white in face
4. Flat shins
5. Legs other than orange, yellow or light willow
6. Squirrel tail
7. In the Silver male, chestnut colouring in the wings.

Breeding Legbars

Legbars bred true to type and colour if due attention was paid to selecting the breeding pen. However, should new blood be needed or someone wishes to remake the breed, the following diagram shows how this can be achieved.

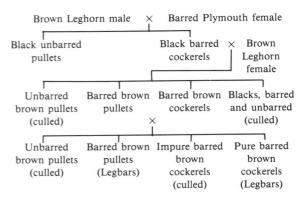

It is of interest that other autosexing breeds could also be remade again using the appropriate parent stock and the above breeding scheme.

To the author's knowledge, there are no Legbar or autosexing breed bantams in this country but there is no reason why they too could not be remade using the appropriate bantam stock. This is perhaps, a project for somebody with an interest in developing new or remaking breeds.

Australorp

Breed History

The Australorp has been proudly seen as our national breed, taking on the world in egg production, but to its disadvantage, was embroiled in an interstate conflict for many years over a suitable Standard. Remarkably, countries importing the breed were able to agree on an acceptable Standard but not so in the country of the breed's origin, until 1949.

The original stock used in the development of the Australorp was imported from England out of the Black Orpington yards of William Cook and Joseph Partington in the period from 1890 to the early 1900s. Local breeders used this stock together with judicious outcrossings of Minorca, White Leghorn and Langshan blood to improve the utility features of the imported Orpingtons. There is even a report of some Plymouth Rock blood also being used. The emphasis of the early breeders was on utility features. At this time, the resulting birds were known as Black Utility Orpingtons.

The origin of the name 'Australorp' seems to be shrouded in as much controversy as the attempts to obtain agreement between the States over a suitable national Standard. The earliest claim to the name was made by one of the poultry fancy's past 'institutions', W. Wallace Scott, before the First World War. Equally as persuasive a claim came in 1919 from Arthur Harwood who suggested that the 'Australian Laying Orpingtons' be named 'Australs'. The letters 'orp' were then suggested as a suffix to denote the major breed in the fowl's development. A further overseas claim to the name came from Britain's W. Powell-Owen who drafted the British Standard for the breed in 1921 following an importation of the 'Australian Utility Black Orpingtons'. It would be a brave person to judge who was right. However, it is certain that the name 'Australorp' was being used in the early 1920s when the breed was launched internationally.

It was the egg laying performance of the breed which attracted world attention when in 1922–23 a team of six hens set a world record of 1857 eggs at an average of 309.5 eggs per hen for a 365 consecutive day trial. It must be remembered that these figures were achieved without the lighting regimes of the modern intensive shed! Such performances had importation orders flooding in from England, United States of America, South Africa, Canada and Mexico. Overseas opinion of the fowls can be gauged by this American commentary following an importation of Australorps in the early 1920s. 'The Australorp is a happy, contented fowl with beautiful glossy plumage bearing lustrous green sheen. Brightness of comb and wattles indicates alertness—yet they are gentle and lovable. These features make them to us the most desirable fowl in the World.' A glowing tribute for any breed!

The breed was standardised and admitted to the American Standard of Perfection in 1929. The egg producing areas of southern California proved to be the stronghold of the breed and it was not long before the Australorp Club of America was set up.

Across the Atlantic in Britain, the breed was also finding a new home, with birds being imported after World War I. On 22 July 1921 'The Austral Orpington Club' was formed and a Standard drawn up. In October 1923, the club was renamed 'The Australorp Club' and the breed assumed the nickname 'Boomerang breed', a reference to the fact that its predecessors were exported from Britain only to be imported in a refined utility form.

It is interesting to note that even South Africa had the breed standardised in the 1920s and that Australorps were playing an important utility role in that country.

Meanwhile, in the breed's country of origin controversy raged over what would be regarded as a suitable Standard for the Australorp. The position was that each State of the Commonwealth had its own Standard but none would give way to accommodate a National Standard. This was not achieved until 1949 at a now famous meeting at the Royal Easter Show of that year.

The catalyst for this was Edwin Hadlington, N.S.W. Department of Agriculture's Chief Government

Poultry Officer, who under the chairmanship of Mr R.R. (Bob) Brown convened a meeting of delegates from all States of the Commonwealth to resolve the issue. It is of interest that the Australian Standard, almost word for word, is that which appears in the current British Poultry Standards, such is the respect for the now famous 1949 meeting.

Any discussion of the Australorp's history would not be complete without some reference to the work of Edwin Hadlington. Prior to the national standardisation of the breed, Hadlington set about his own breeding program. He was, at the time, not happy with the direction the Utility Orpington was heading. His breeding program began with a visit to Athol Giles' famous 'Bonaventure' Poultry Stud at Mt Druitt where he selected twelve close feathered, good headed Black Orpington cockerels from a group of sixty. These were taken to the State government's Seven Hills Research Station where they were mated with selected females on the farm. Once this strain was established, they were crossed with Langshan blood. The selected Langshan strain featured a large quick eye, full open faces and

minimal leg feathering. The effect was to further tighten the feather and improve the up to then poor body sheen.

Further to this, birds from N.S.W. South Coast and Gunning (N.S.W.) breeders were introduced into the Seven Hills flocks to further improve the stock. What was developed was an almost national stud from which producers and fanciers alike could select desirable stock.

It is unfortunate that this flock no longer exists. However, the breed does have strong support and classes of over thirty birds can be seen at important shows throughout the country.

Perhaps the breed's greatest tribute was that in 1962 at the World Scientific Poultry Congress an Australorp male was selected as the Congress motif on stamps, letterheads and badges, as a symbol of Australia's contribution to the poultry world.

As to who was responsible for or how Australorp bantams appeared is unknown to the author, but they have been exhibited for many years.

Positive Features to Look For

Important features of type to be looked for are:

1. Alert, active and expressive birds.
2. The lines of the bird should follow the graceful lines of the Orpington but the body is longer than it is deep. This is a breed of curves not angles. The breast should be well rounded. The breastbone should be straight with a fine point slightly sloping downwards.
3. The feather underline should not protrude below the hocks.
4. The body should be well balanced with the legs set well apart and not giving the appearance of being simply stuck under the bird. The bird should appear upstanding but balanced.
5. The Australorp is a breed which should handle 'larger than it looks'.

6. In the head, look for a good black eye, a clean red face in both sexes and small thin lobes of sound red colour.
7. Particular attention must be placed on the tail which must be full and, in appearance, rise gradually from the saddle in an unbroken line. In the male, the sickles must gracefully curve to finish the bird. Make sure the feather is sound with no soft quills. A solid black quill should be seen.
8. The following weights are preferred:
   Cock 4 kg
   Cockerel 3.6 kg
   Hen 3.2 kg
   Pullet 2.7 kg

Colourwise, the fowls should be a solid black with a beetle green sheen.

Large Australorp female. Camden Poultry Club Annual 1988. Owner: B. Butler.

Australorp large fowl male. 1988 Sydney Royal Easter Show. Owner: Bill Trimmer.

Australorp bantam female. Fairfield Poultry Club Autumn Show 1988. Owner: Mel Pearson.

Australorp bantam male. 1988 Sydney Royal Easter Show. Owner: R.E. Forrest.

The plumage must not be too loose, that is, of a medium closeness but not as close or hard as in game fowls.

Negative Features to Avoid

The Australorp Standard requires that birds with the following faults should be passed, that is, not considered for further judging:

1. Underweight fowls
2. Birds with wry tails, roach back and crooked breastbones
3. In the feet and shanks, the following should be passed: yellow or willow colouring on shanks or feet; webbed feet or feathering on shanks or feet, which stems from Langshan blood
4. Unacceptable eyes include red, yellow or pearl eyes, or grey iris of the eye
5. Side sprigs on combs
6. Split, twisted or slipped wing.

The Australorp standard also regards the following as serious defects:

1. Birds showing red, yellow or white in feathers
2. Permanent white in the ear lobes which is a throw back to Minorca blood.

Other faults that do appear and need to be culled out include:

1. Birds with angular appearance as opposite to the 'bird of curves'
2. Excessive fluff particularly over the hocks. Such birds are said to show 'blousiness' or more bluntly are referred to as 'feather dusters'
3. High and swinging tails. Likewise, birds showing long streamy sickles. This is associated with the infusion of Minorca blood.
4. Dull body colour. This can be picked up at an early age in youngsters. Birds that will grow into dull black or grey adults have dull chicken feathers and they generally show up in the breast first. Again in the growing stages, such feathers come through on the back. They do not show the desired rich green sheen.
5. Cull out birds with purple or blue sheen. Fanaticism with breeding excessively green birds can lead to purple sheen. Male birds carrying the

tendency to purple show it in their tail furnishings.

6. Avoid facial coarseness
7. Gypsy faced (mulberry) birds as these go with purple sheen in the feathering
8. Excessively heavy boned birds.

Breeding Hints

1. The breeding priorities for Australorps are:
   Firstly, type
   Secondly, size
   Thirdly, stamina.

2. Eye colour has a sex linkage pattern meaning that the dense eye colour of the father is passed on to his daughters and likewise, if a female possesses light eye colour, it will be passed on to her sons.

3. The quills of the tail feathers should be carefully looked at, especially in the males. Birds that show soft quills should be culled and not used in the breeding pen. Breeding from soft quilled birds produces birds taking up to 12 months to tail. The slowly produced feather buds and sheaves are often the target for growing youngsters that are disposed to feather picking, as the bared tail area seems to attract such birds. This further aggravates the tailing of the Australorp youngster. Without a tail, a young male's potential can never be accurately assessed as the bird always appears out of balance.

4. Whilst on the topic of feathering, there is evidence that two types of feathering of chickens exist, that is, fast and slow featherers. The feature is also sex linked. Later in life, the fast feathering strains have the advantage that they moult quicker than the slow feathering strains. The rate of feathering trait has the following features:

(a) all females are pure to the feather type they exhibit
(b) fast feathering males are pure to type but slow feathering males carry one gene for each trait
(c) the male determines the rate of feathering in the female offspring
(d) the female always transmits the fast feathering trait to her sons
(e) rate of feathering is responsive to selection from a previously unselected flock, that is, there is a high selection differential. This means that this feature can be built into a strain of Australorps.

Some skill and practice is needed to identify chickens of each strain but the two strains show the following characteristics:

(a) Soon after hatching, the chicks of the fast feathering strain have primary wing feathers up to one centimetre long arranged in an alternate fashion along the wing.

The difference between the short and long feathers is about half a centimetre. The slow feathering strain have even primary feather growth showing a variation of as little as a quarter of a centimetre.

(b) The differences between the two strains becomes much more apparent at two weeks of age. The fast feathering strain shows the following features. The primary feathers reach the end of the body, the secondary feathers are well grown and the tail feathers are about one centimetre long. Another important factor exhibited is the more desirable broader feather. On the other hand, the slow feathering strain shows no tail feathers, primary feathers only half to three quarters of the body length and poorly developed secondary feathers.

5. Turning to colour breeding in the Australorp, we have a diversity of opinion. Mastering this feature requires a constant look-out for certain warning signs. Over-correction, however, can lead to other problems. The following general comments can be made:

(a) There is a tendency for two brilliantly sheened parents to throw purple sheen in their offspring.
(b) Better results are generally achieved by using birds of good feather quality, not over-sheened but showing a definite tendency to green.
(c) The eye and shank colour appear to be linked.

Some people go as far as to advocate the following:

(a) Look for a male with a little bright red in the neck hackle. An excess of red, however, will produce cockerels with red on the wing bow.
(b) To correct dull body colour, choose sound hens and mate them to a cockerel with brilliantly green top colour but just a little lighter in under colour.

The author is of the opinion that these techniques would work but there is a danger of the breeder moving into a 'double mating' situation. The Australorp has been fortunate up to date in that exhibition stock can be produced from a single pen. This has been an attraction for many fanciers. Careful consideration would be in order before such drastic breeding strategies were taken.

Before leaving the Australorp, it may be of interest to know that white Australorp sports have been produced but the colour, to the author's knowledge, has never taken off.

Barnevelder

Breed History

The initial appeal of this breed has always been the large coffee brown eggs the Barnevelder female lays. The breed was developed from Orpington, Cochin, Brahma and Croad Langshan blood around the Dutch town of Barneveld from which the breed subsequently took its name. The Dutch were quick to recognise the lucrative market in London for large brown eggs for which the British were prepared to pay a premium price. In turn, the British imported the breed in 1921 and set about standardising it. It is of interest to note that there are differences between the British Standard and the Dutch Standard today.

Barnevelders were used in the early work on sex linkage, behaving as a 'gold' when a male is crossed with a 'silver' female such as the Light Sussex or the Columbian Wyandotte. This meant that the sex of the chickens could be determined by the colour of their down at hatching.

The British Poultry Standards recognise four colours, the Black, Double Laced, Partridge and Silver. However, most interest lies in the Double Laced variety, both here and in Great Britain. The breed of recent times is showing a revival of interest, particularly in the hobby farming belt of southern Australia.

A bantam version of the fowl does exist overseas but to the author's knowledge does not exist in this country. There are some breeders currently working to bantamise this breed but to date most are too large and lack the colouring of the large fowls.

Positive Features to Look For *(colour plates p.83)*

Physically the breed is classed as a heavy breed with mature males weighing up to 3.6 kg and mature females 3.2 kg. The stance should be keen and alert. The fowl should have a compact body, showing depth through the body, a short back and close feathering. The hocks should be clearly visible in the underline of the fowl. The shanks well spaced under the body when viewed from in front or from the rear. The beak and shanks should be yellow in colour.

Since the main interest is in the Double Laced variety, the following remarks apply to it. When looking at the hackle feather of the male, it should be black with a red brown quill and surrounded by a thin red brown edge. The covert feathers of the back are

red brown with a wide black lacing around the outside. Grey fluff is present at the base of the tail.

The female, on the other hand, when first viewed, appears very similar to the Indian Game female with well rounded lacing all over, particularly the lower part of the body.

Negative Features to Avoid

Physical features to avoid include:

1. Undersized birds
2. Narrow bodied birds when viewed from the front or rear
3. Coarse headed fowls with any attendant problems such as double serrations
4. Light or green eyes
5. Black or very dark legs
6. Feather stubs in shanks which are probably throwbacks to the original breeds used in producing the Barnevelder.

Colour features to avoid include:

1. Triangular or pointed lacing as the lacing must follow the shape of the rounded covert feather
2. Yellow or yellowish shades in the ground colour instead of the desired red brown
3. White in the under colour, wings or tails.

Breeding Hints

The breeding priorities can be listed as follows:

1. Type: This holds true for all breeds.
2. Size: The breed is a heavy breed and high priority must be given to size and the selection of a heavy breeding female.
3. Egg laying ability and egg colour: Since the breed was developed for supplying their famous dark brown eggs for sale, maintenance of these features is essential.
4. Colour: Correct shade of ground colour in the male is necessary or the ground colour in the resulting females will be weak.

Often, when breeding for lacing becomes a preoccupation, the type and size suffer. Breed type must remain the top priority if the breed is to remain recognisable.

Belgian Bantams

Breed History

This is an ancient group of bantams with no large fowl equivalent. As the name implies, they originated in Belgium. Two breeds of Belgian bantams that have caught the attention of poultry fanciers are the Barbu d'Anvers and the Barbu d'Uccle. Both are extremely popular in Great Britain, Europe and the USA. The Barbu d'Anvers is numerically most popular in Australia but a group of keen fanciers are building up the small numbers of the Barbu d'Uccle.

The appeal overseas is the type and seemingly endless range of colours, some being extremely unusual and challenging to breed. In Australia, their distinctive type has caught the imagination of a keen band of followers who seem to find little trouble in gaining new 'converts' to the breed. The Lavender variety seems to have the strongest following.

The Barbu d'Anvers, also called the Bearded Antwerp, features a rose comb and clean legs. The Barbu d'Uccle, coming under the name Bearded d'Uccle, is single combed and has feathered legs. The Bearded d'Uccles were produced in the early 1900s by crossing Bearded d'Anvers with Booted bantams and following up with a selective breeding program. The Barbu d'Anvers on the other hand have been around since the sixteenth century.

Positive Features to Look For *(colour plates pp.84–5)*

Since the Barbu d'Anvers are the principal Belgian breed in this country, the following remarks apply to them:

1. Carriage: Proud, upright, a distinct jaunty type
2. Body: Short and broad
3. Breast: Arched and carried well up
4. Back: Very short, sloping down to the tail
5. Tail: Almost upright around 75°
   Sickles short, narrow and pointed upwards 2–5 cm past the ends of the tail feathers with the rest fanned
6. Neck: Arched and of medium length
7. Wings: Sloping towards the ground and of medium length
8. Eyes: Large
9. Beak: Short and strong
10. Comb: Rose comb, leader clear of the back of the neck
11. Beard: The beard is made up of three areas of feathers, two on the sides of the beak and one underneath the beak
    The side areas turn horizontally backwards from the sides of the beak and those of the central area turn downwards
    The overall impression is that the bird is wearing a collar of feathers
    The wattles and ear lobes, though very small and poorly developed, are hidden by this 'beard' of feathers.

Negative Features to Avoid

Reject any features which destroy the breed's type mentioned in the foregoing section. These include:

1. Long backed birds
2. Low tail carriage
3. Leader on comb pointing upwards and away
4. Feathered shanks
5. White ear lobes
6. Squirrel or wry tail
7. Obvious ear lobes and wattles through the beard
8. Yellow legs or skin
9. Single comb
10. Dubbed wattles.

Breeding Hints

The breed Barbu d'Anvers appears to be reasonably fertile although the chickens seem very delicate and would be best reared in groups by themselves away from other larger chickens. The breed, from all reports, breeds reasonably true to type and colour provided due attention is paid to selecting the breeding pens.

Brahma

Breed History

This breed, to the author's knowledge, has disappeared from the Australian poultry scene. However, it has been included in the remote hope that maybe somewhere there are still some in existence that have escaped the eye of the poultry fraternity and, through their recognition, could be brought back to people's attention. Alternatively, with the opening up of quarantine facilities, some people may see their way to reintroduce the breed into this country. The breed has also been included because of its influence on many other breeds we know today.

The original stock used to produce this breed was imported from India into the USA in 1846 and was subsequently exported to Great Britain sometime later. The variability of the stock caused much conflict between those concerned in the two respective countries. To add to the conflict, stock which was sold as 'Indian' in origin had in fact been imported from China.

The Americans were the first to standardise the breed from what was known as 'Chittagong' stock, a Malay-Cochin cross. The type of fowl desired was a long legged, tight feathered, active, good foraging bird with little leg feather. This was the bird that played an important role in the development of the Rhode Island Red and Wyandotte breeds.

Across the Atlantic, in Great Britain, more emphasis was being placed on the development of leg feathers and it is this abundance of feathers that distinguishes the two types.

In Australia it is believed the breed first arrived here in the early 1850s, making sufficient impact that by the 1860s Chittagong type fowls were being imported directly from India to improve the utility properties of the Dorking, Old English and Hamburgh breeds. The Australian Brahma was developed along British lines. Initially, it came to general attention as a good layer and table bird. However, the fancier then set about exaggerating feather features at the expense of the utility properties. Many claimed that the end product was an excellent example of the fancier's art, a handsome and majestic fowl. Others claimed that it was a case of exaggerating useless, over abundant feathering at the expense of the breed's utility properties, ultimately leading to the breed's downfall.

Brahmas existed in two forms in Australia. Firstly the Light which had a Columbian pattern, and secondly the Dark with a feather pattern similar to the Silver Pencilled Wyandotte.

Brahmas were used by utility poultry farmers for crossing with Game, Malay, Houdans and Dorkings to produce table fowls. The Brahma-Malay cross was particularly favoured because of the size of the offspring produced. When crossed with Spanish and Minorca fowls, good egg layers were produced.

However, on the show scene the breed gradually fell into disfavour and subsequent apparent extinction as indicated by a report in the *Poultry* newspaper of 18 July 1959 which says, 'So far as we are aware, the last Light Brahmas to be exhibited in N.S.W. were benched at the Sydney Royal Show in 1940'. It is hoped that the breed finds some new friends in the near future.

The author has uncovered a breeder, George Childs who has used information from the poultry literature of Entwistle to re-make a line of Brahma bantams. Photographs of his efforts have been included on page 84. George is to be congratulated for his efforts.

Campine

Breed History

The Campine breed has been known in Belgium and France since the sixteenth century. It has its origins in the La Campine country of Belgium where it is also known as the Braekel. In its place of origin it has been kept as a quick grower and producer of fine white eggs.

The breed came to Great Britain when the Huguenot refugees fled religious persecution in their homeland, so the breed has been present there for a long period of time. The Campine was 'rediscovered' in 1899 by poultry fanciers, enjoying a deal of popularity particularly following its involvement in the development of the first autosexing breed, the Cambar, a cross between the Campine and the Barred Plymouth Rock.

Features that attract people to the breed are its active, sprightly nature, non broodiness, foraging ability, quick maturity, and laying persistency which extends into the third or fourth laying season, but the most obvious attraction is the plumage. Both male and female have a barred body and clear neck hackle of gold or silver presenting an eye catching picture in any company. As indicated, the breed comes in two varieties, gold and silver, referring to the ground colour on which the barring is placed. The male is different from other breeds in that the saddle is barred like the female so that both sexes present the same colour pattern.

When the breed arrived in Australia is not known, but by the 1920s, Campine breeders had their own specialist breed club which by all accounts was well supported. At the time people were attracted by its egg laying ability, white eggs and plumage beauty. Today, the breed would have to be regarded as in the 'rare breeds class'. It would be well served by some extra supporters, particularly those interested in preserving a little genetic heritage for future generations.

No bantams are standardised in this breed.

Positive Features to Look For

The breed type we are looking for can be summarised as follows:

1. Carriage: Alert and graceful
2. Body: Broad, compact and narrowing at the tail
3. Breast: Prominent and well rounded
4. Back: Rather long
5. Wings: Large and carried well up

6. Tail: Tail carried well out at about 45° to the backline.
   The male should have plenty of broad sickle feathers
7. Neck: Arched
8. Comb: The male's comb needs to be well cut, carried well out and clear of the neck.
   The female's comb should curve neatly to one side

9. Beak: Short and horn coloured
10. Eye: Bright, and dark brown with a black pupil
11. Lobes: Almond shaped and white in colour
12. Face, comb, wattles: Smooth and bright red in colour
13. Legs and feet: Medium length, leaden blue in colour
14. Weights: Desired weight for a male is around 2.73 kg and for the female 2.27 kg.

Turning our attention to the plumage:

1. The plumage in general should be close fitting.

2. The head and neck hackle should be either a rich gold or pure white depending on the variety.

3. The barring needs to be a clean transverse bar across the feather, around three times as wide as the ground colour bar. To look its best, the barring needs to appear to form rings across the body. To do this, some of the barring on the breast and underparts needs to be straight or have a very slight curve.

Looking at individual feathers, the barring should be a distinct clean beetle green bar across the feather. The end of the feather should have a tip of the ground colour.

When assessing the barring, it is important to look for evenness of barring across the bird in all parts not in just one or two parts of the body.

Negative Features to Avoid

Physical or type features to avoid include:

1. Squirrel or wry tail
2. Feathered shanks or any showing other than the required leaden blue
3. Side sprigs in the comb
4. White in the face
5. Red eyed birds.

Plumage faults that need to be looked at include:

1. Barring which is equal width with the ground colour
2. A blue grey instead of the desired beetle green barring
3. 'Mossy' barring which is most likely to occur in the wing feathers or back plumage
4. 'Horse shoe' barring on the breast
5. Black spots, uneven markings or white patches on the breast
6. Pencilling in the ground colour
7. Ground colour which is yellow and not the required gold in that variety.

Breeding Hints

There is no need for double mating in this breed although some selected breeding stock may need to be kept to breed out weaknesses that occur.

Breeding priorities that need to be looked at before paying too much fanatical attention to barring are firstly, size, secondly, type and thirdly, fullness of front. It is important to remember that the Campine is an egg laying breed so this feature should not be allowed to slip. By keeping up the size, the fowl remains active and there is a greater chance of maintaining the desired rounded breasts.

'Mossiness' in the barring can be bred out of a strain if particular attention is paid to selecting the male to head the pen. He should be a 'dark' male with sound beetle green barring. Likewise, 'blue-grey' barring can be bred out using a similar male.

The main breeding challenges with plumage are in achieving good hackle and well barred breasts, but these can be achieved with patient breeding.

Dorking

Breed History

Dorkings are another breed to be put into the 'rare breeds' group in this country, although this was not always the case. They were one of the earliest breeds imported into Australia from Great Britain probably because of their excellent table features. Royal Agricultural Society of N.S.W. records show that classes were provided for Dorkings at the 1867 Show and that these were well supported.

Later, Dorkings were used to cross with other imported breeds such as Indian Game, Sussex and Faverolles to produce top quality table fowls.

The breed appeals with its characteristic low carriage, stately disposition and docile nature. They have surprisingly long wings for such a big bird and can fly well. As a table bird they appealed because of their large size at an early age, fineness of bone, relatively small amount of offal, quantity of breast meat, whiteness and quality of meat.

Dorkings have an ancient history with their ancestral stock claimed to stretch back to the Roman invasion of England. It is on record that Columella recorded the presence of a five toed fowl of good table quality being kept by the early Britons.

During the middle of the nineteenth century, the breed underwent a deal of refinement. Lewis Wright, noted poultry expert of the time, mentions the crossing of the Red Dorking fowl with Silver Duckwing Game. The table properties improved but in the preoccupation with feather colour, the laying capabilities were reduced such that the Dorking is not now regarded as a noted layer.

Today the British Poultry Standards recognises five colours, Silver Grey, Cuckoo, Dark, Red and White, the Silver Grey being the most popular. Small numbers of Dorking bantams exist in this country.

Positive Features to Look For *(colour plate p.86)*

Looking at the distinctive physical features of the breed that are present in a good Dorking type:

1. Body: It is carried in a stately fashion
   Massive, long, deep but compact
   Carried well forward of thighs
2. Breast: Deep, well rounded
   A long straight keel

3. Back: Broad and moderately long
4. Weight: Up to 6.35 kg for the male
   Up to 4.55 kg for the female
5. Legs: Short and strong
6. Toes: Front three toes long and straight
   Fourth (normal rear toe) straight back to give the fowl good balance

Fifth toe must be well developed, above the fourth toe, pointing upwards preferably at an angle of 45°.

Remember the fifth toe is a distinguishing feature of this breed so it must be present.

Since the breed is such a heavy one, the feet play an important part in the bird's ability to walk about and forage. Good, well structured feet under a large bird are essential.

The remarks on colour to be made here only apply to the Silver Grey variety, as to the author's knowledge the others do not exist in this country.

Male colour
The neck and saddle hackle feathers are a clear silver white. A steely blue black bar is present across the wing. The wing feathers are black with a white outside edging. The rest of the plumage black.

Female colour
The body colour is a clear silver grey with fine grey pencilling giving the bird a soft grey appearance. More heavily pencilled birds appear a darker shade of grey. The head and neck are silver white, striped with black. The breast needs to be a red shade of salmon toning off to ash grey at the thighs. The main tail feathers black.

In both male and female, the following colours should be present:

Eyes: Bold red
Beak: White or horn
Comb, face, wattles, ear lobes: Red
Legs: white

Negative Features to Avoid

When turning our attention to features that need to be avoided, look for the following:

1. Birds with crooked breastbones and rounded instead of straight backs
2. Birds with feet problems or 'tender feet' such as 'bumble foot', swollen toes and crooked toes
3. Coarse headed birds
4. Excessive comb development
5. Flat shins
6. Yellowness across the back and hackles which is regarded as a serious defect
7. White in the black breast or tail of the male
8. White in the ear lobes instead of the required red
9. Often hens with good top colour have a pale breast or hens with good breast colour often have a harsh top colour.

Breeding Hints

1. Unfortunately for the Silver Grey Dorking, the number of points allocated to colour is more than that for the breed's main distinguishing feature, its size. The temptation would be to place size as a breeding priority second to colour. This should be avoided as a Dorking without size is simply not a true Dorking. Size should be the number one breeding priority.

2. As male birds age, they often have a tendency to develop white patches or 'grizzling' on the black feathers of the thighs. Provided the bird was initially sound in this area it can still be used successfully in the breeding pen, but bearing in mind that this problem will also occur in the male progeny at a later date. It would therefore be best to keep this to a minimum.

3. Attention should also be paid to leg length, bearing in mind that this is a low stationed breed. Over emphasis on this feature will have birds being produced that are too low to the ground. Thus, a pen of hens short in the legs is best mated to a male a little up on his legs. Likewise, a pen of hens up on their legs can be successfully mated to an exhibition type male to balance out leg length.

Faverolles

Breed History

Faverolles were first imported into Australia in the 1920s by a Madam Masseran who had spent time working in the area of France where the breed was developed. At the time of their importation, Madam's husband was the chef at the exclusive Melbourne Club and she saw the opportunity to meet the exclusive tastes of the Club's patrons by providing the highly sought after flesh of the Faverolles. This she did. In fact she later went on to supply overseas orders from the British royalty and aristocracy.

Walter Hawker of South Australia became the Faverolles Club's first president and under his chairmanship the breed enjoyed a strong following up to World War II, mainly due to the excellent table qualities of the birds. Since then, breed numbers have declined, but Faverolles do have some supporters in southern N.S.W., Victoria and South Australia. No bantams seem to be raised in this country.

The breed has its origins in the Eure et Loir area of northern France where it took the name of the Faverolle village. Originally, it was developed as a dual purpose breed, producing heavy table fowls and winter layers of tinted eggs. The breed was developed by crossing the Dorking with the bearded and muffed Houdan. Cochin blood was also used in the original development.

The breed was imported into Great Britain in the 1880s where it was crossed with many breeds including Sussex, Orpington, Rhode Island Red, and Indian Game to produce large crossbred meat chickens with white flesh of excellent flavour. The Salmon and Ermine colours were the only colours imported into Great Britain from France, but other colours were developed by reusing Brahma and Cochin blood. Today the British Poultry Standards recognises the following colours: Black, Blue, Buff, Ermine, Salmon and White. Bantams have been developed in a number of these colours and standardised.

Only the Salmon and White are recognised in the USA. Bantams are recognised there.

Positive Features to Look For

Physical type features that need to be looked at are as follows, remembering that this breed is basically a table breed so it is those features which are important.

1. Body: Size is the first priority
    The body should be deep and thick, what the old timers called 'cloddy'
    Side view of the body should be rectangular.
2. Breast: A wide, deep, full breasted, long flat keeled bird is needed
3. Back: Straight
4. Weight: The male up to 4.54 kg and the female up to 4.09 kg
5. Tail: Moderately long, carried fairly upright and made up of wide tail feathers
6. Comb: Single, 4 to 6 evenly cut serrations
7. Legs: Relatively short but not to the shortness of the Dorking
8. Toes: This is a five toed breed so the fifth toe must be located above the fourth (normal rear toe) and inclined at 45°
9. Muffs and beards: Compact and give the impression of a full face beard.

The legs and skin need to be white in colour.

Since the Salmon variety is the main variety represented in this country, the following remarks apply to it.

The male
Neck: Straw coloured
Back and wing bars: Bright cherry mahogany
Wings: Black with white outer edge on the secondary feathers so when closed forms a white bar
Beard and muffs: Black
Rest of body: Black.

The female
Neck: Straw coloured, striped with a darker shade but not black
Beard and muffs: Creamy white
Breast, thighs, fluff: Cream
Rest of body: Wheaten (reddish) brown.

The beak colour is white in both sexes.

Negative Features to Avoid

In summary, the features to avoid are in essence the opposite of the desired features. Two faults though, need close scrutiny: firstly, thin, cut away fronted birds and secondly, those with poor feet.

Breeding Hints

1. Since Faverolles are table birds, the first priority is size. It is important to select oblong, deep bodied hens to ensure that the desired size is passed on to the next generation.

2. Whilst the Standard allocates 15 points for beards and muffs, care must be taken not to become overly preoccupied with this feature at the expense of size. To obtain the desired muffs and beards, balanced breeding may be required. If an otherwise good bird is lacking, mate it with a fowl that is slightly overdone.

3. The same balancing approach can be applied to the hock feather. Overdone fowls should be mated with sparsely feathered birds.

4. The late Harry Maude gives the following advice on colour. Other colours have been included in case they appear on the local scene by way of importation or local breeder development.

Salmon: Use the richest topped, sound black breasted cockerel to the lightest topped hens, and vice versa.

White: Use pure white hens even if they show a black feather or two as they tend to throw the whitest of offspring. A male that shows a yellow tinge or straw colouring across the back can safely be used with pure white females without too many problems.

Blue: Light blue to dark blue matings are preferred. A well laced female is best mated to a dark cockerel.

Black: Use the densest black birds with colour running to the skin. Difficult to avoid purpling in this variety.

Buff: Mate dark golden buff pairs with the least peppering in flights and tails.

Before concluding breeding, it must be pointed out that care is needed with the conditions the fowls are kept under if the beards and muffs are to be kept in top condition. The birds would be best protected from the weather in a semi intensive situation.

Frizzle

Breed History

The Frizzle is believed to be of Asiatic origin and is regarded as a purely ornamental or exhibition fowl with attention catching feathers that curl backwards and upwards towards the head. Such ornate plumage necessitates careful washing and preparation for exhibition.

Darwin noted that 'frizzled' fowls were seen in India where they were also known as 'Caffie' fowls. 'Frizzled' fowls were also noticed by other observers to be native to Java, Japan and other parts of eastern Asia.

Whether they are really a distinct breed or not is open to much debate, as the American Bantam Standard does not regard them as a breed but as 'frizzled' versions of other breeds. On the other hand,

the Frizzle was standardised in Great Britain in the late nineteenth century, being regarded as a breed in both large and bantam fowls. The bantams are represented in that country in much larger numbers with the large fowl version almost non-existent. In Australia, the breed is quite well represented in the bantam form with a number of avid breeders to keep the breed alive and well. Frizzles are mainly seen in Black, Blue, Buff and White but other colours based on Old English Game colours have been developed.

Because of their frizzled plumage, they are regarded as good broodies. Their plumage also makes them poor flyers and easy to restrain.

Positive Features to Look For

Before looking at the breed's obvious attraction, the frizzled feathers, it is important to establish the type of bird needed to carry those feathers.

1. Carriage: Confident and cocky
2. Head: Fine
3. Back: Broad but short
4. Breast: Well rounded and carried forward
5. Wings: Long with a low carriage
6. Tail: Carried confidently, erect, broad with plenty of wide feathers.

Turning our attention to the plumage where 45% of

the points are allocated to curl and feather quality or crispness of curl:

1. The individual feather should be broad, and curled backwards towards the head. The curl should be close and even. Each feather must also possess length.

2. The breast needs to be made up of many curled feathers to give the fowl the necessary wide appearance there.

3. In the male, the neck hackle area is referred to as the mane. It is important to have a pronounced curl towards the head and plenty of it.

Pair of Frizzle bantams. Fairfield Poultry Club Autumn Show 1988. Owner: Mel Pearson.

White Frizzle bantam male. 1988 Sydney Royal Easter Show. Owner: J. Crabb.

Negative Features to Avoid

These will be divided into general physical faults and plumage problems to avoid.

1. General physical faults include:

(a) Narrow bodies
(b) Short necks
(c) Overly long or large tails
(d) Large coarse combs
(e) White in lobes
(f) Weak leg colour.

2. Plumage problems to be on the look out for include:

(a) Narrowness of feather
(b) Wispy plumage
(c) Crosses with other breeds which can be detected by looking at the upper breast feathers and on the back, where the feathers appear much shorter than they should be in a truly bred Frizzle.

Breeding Hints

1. Exhibition Frizzle fowls do not breed true to their feather type. Over reasonably large numbers of offspring, the following percentages of feather types appear: 25% normal feathers, 50% exhibition feathering and 25% that are called 'woolly' or 'curly' or 'stringy curlies'. The last group are generally slow maturers and should be culled at the earliest opportunity. Another feature is that they cannot fly.

2. Persistent mating of exhibition Frizzles also leads to a narrowing of the feather, so an occasional outcross is necessary to maintain feather width. This is usually achieved by mating an exhibition male to a plain feathered female of Frizzle breeding. The selection must be based on feather width in this female.

For those who enjoy purely ornamental poultry, the Frizzle represents a rewarding challenge.

Hamburgh

Breed History

Today the Hamburgh is regarded as an exhibition fowl. However, most poultry experts agree that it is one of the oldest breeds of poultry still in existence, and the ancestors of the Spangled varieties can be traced back three hundred years in parts of England. The name Hamburgh was given to the breed in the early 1800s as the original stock was imported from Holland via Hamburgh in Germany. Poultry breeders of the time set about standardising the breed in Great Britain and developing new varieties from their own native Spangled stock, to arrive at the Hamburgh as we know it today.

Originally the breed was imported into Great Britain as a laying fowl and by all historical accounts performed well, but it was as an exhibition fowl that the Hamburgh made its greatest impact.

In Great Britain the breed has been standardised in Black, Gold Pencilled, Silver Pencilled, Gold Spangled and Silver Spangled varieties. Bantams are also recognised. In the USA the White variety is also accepted by their Standards.

The breed was one of the first pure breeds to arrive in this country, being catered for in the Royal Agricultural Society of N.S.W.'s show in 1867 where classes for only Dorking, Hamburgh and Game were provided.

Teams of Hamburghs took part in early Hawkesbury laying trials, performing reasonably well but no match for the developing commercial strains of Leghorns, Australorps and Langshans. Their commercial future was limited as they were regarded as too flighty and needed double mating to be successfully bred to type.

Today, Australian breeders believe their birds are of world class with an active breed club based in Victoria strongly promoting the breed. Some Hamburgh bantams also exist in that State.

Positive Features to Look For

Here we shall look at physical features and look at colour features in the section on breeding.

1. The bird needs to be active, bold and alert
2. Body: Medium length, rounded and narrowing towards the tail from reasonably wide shoulders
3. Neck: Medium length with lots of hackle feathers in the male
4. Breast: Rounded and full
5. Wings: Large, strong and carried well up
6. Tail: Feathers broad
   In the male the sickle feathers need to be large and well curved, surrounded by plenty of furnishings
7. Legs: Medium length and leaden blue in colour
8. Head: Should support a rose comb with plenty of workings on top, a well developed tapering leader moving out behind the head and showing no tendency to droop
   A pair of white, evenly shaped lobes
   A sound red face

Silver Spangled Hamburgh large fowl female. 1988 Sydney Royal Easter Show. Owners: R. & D. Perkinson.

Silver Spangled Hamburgh male. Camden Poultry Club Annual 1988. Owners: R. & J. Perkinson. Tail not fully out following the autumn moult.

9. Beak: Horn coloured
10. Weights: The male should be around 2.27 kg and the female 1.82 kg.

Negative Features to Avoid

The following need to be looked for and avoided or culled out:

1. Crooked breast bones
2. Loose, fluffy plumage
3. White in the face
4. Squirrel tails
5. Combs not sitting squarely on top of the skull
6. Combs with bent leaders or ones with nobs protruding off them
7. In the Spangled varieties, look for 'mossy' ground colour (black markings on the ground colour). Also look for heavy markings which overlap and lose their distinct shape
8. In the Pencilled varieties, look under the throat and at the top of the breast of the female as it may be spotted rather than pencilled. Look for 'ticking' rather than pencilling in the neck hackle.

Breeding Hints

All Hamburgh varieties need double mating to successfully produce exhibition stock of the desired type.

1. Black

Require double mating to maintain the brilliant beetle green sheen which is a feature of this variety. Successive sheen to sheen matings produce progeny which has a tendency to go purple. A sheen to 'flat' black mating overcomes this.

2. Spangled

Whether the variety is Silver or Gold, the same underlying pattern applies. The difference between the two varieties is the ground colour, which in the case of the Silver Spangled variety is silvery white while in the Gold Spangled variety it is rich bright bay or mahogany.

The spangles and feather tips are a rich green–black in colour. The body spangles are 'half moon' shaped on the body, but males have smaller dagger shaped markings on their hackles, shoulders and backs. Across the wing bows should be two rows of large, round black spangles running parallel around the wing. A large round spangle should tip each secondary wing feather. The female does not have any dagger shaped markings on the back, instead, they are rounded to suit the natural rounded feather of the female.

(a) The cockerel breeding pen would be made up as follows. An exhibition male with precise dagger markings and a long flowing tail mated to a female with pear-shaped spangles, white under colour and top two tail feathers tending towards sickle-like feathers.

(b) The pullet breeding pen would be made up of an exhibition type female with large spangles and black under colour mated to a male similar in appearance to the female, with plenty of body spangles and a short tail.

3. Pencilled

As with the Spangled variety, the Gold and Silver Pencilled feather pattern remains the same with only the ground colour being different: the Silver has a silvery white ground colour and the Gold a bright red bay or golden chestnut ground colour. The male has a black tail, sickles and coverts narrowly laced with gold or silver depending on the variety. In the female, all other feathers apart from the neck hackle are evenly pencilled with fine parallel lines of lustrous black across the feather. To breed Pencilled Hamburghs two pens are required. Included below are two different approaches aimed at achieving exhibition males and females.

(a) Firstly, matings as advocated by the late Athol Giles. The cockerel breeding pen would be made up of a male full fronted, big tailed, thick at the base, with sickles and coverts laced with a gold edging. The cockerel breeding female would need to be full fronted, on the large side, coarsely marked and with the top two feathers of the tail sickled. He advised that it would be important to look carefully at the female's brothers to make sure they were the desired exhibition type. In the pullet breeding pen, use the best pencilled female with evenness of ground colour, avoiding those with pale wing bows and dark bodies. To the selected female mate a male from a pullet breeding line that is almost hen feathered, with plenty of barring on the wings.

(b) Secondly, an approach offered by English breeders. The cockerel breeding pen is made up of a male with no pencilling on the body, black wings and a black tail with lacing (Gold or Silver). The female should have solid black wings, a dark tail and very mossy pencilling. On the other hand, the pullet breeding pen is made up of an exhibition female mated to a male with similar pencilling on the body and wings, and not much black in the neck hackle.

It appears from the above that Hamburghs are a specialist breed and any potential breeder must be prepared to keep accurate records of his stock, as well as keeping the pullet and cockerel breeding strains well separated. For those who enjoy a challenging exhibition fowl the Hamburgh could be the breed for you.

Indian Game

Breed History

Despite the breed's name, these fowls were not native to India but in fact bred and developed around Cornwall in England from Aseel, Old English Game and Malay breeds. In the USA they are known as Cornish. The breed's founders were looking to develop a bird with large amounts of breast and thigh meat. From the first glance at this breed, these points become obvious.

The British Poultry Standards recognise Indian Game as Cornish (Dark) with black lacing and Jubilee with white lacing. In the USA, Indians (Cornish there) are standardised in Dark, White, White Laced Red (Jubilee) and Buff. In Australia, a Blue Laced variety has been developed.

Bantams of the breed exist in all countries but often suffer in the show situation when being considered for major awards as their size often goes against them and, to the uninitiated, they appear the size of a small large fowl, not a bantam.

Indian Game were first imported into Australia in the 1890s but were of a different type from the super wide version we see today. The breed enjoys a stalwart following in this country and is served by a keen breed club.

Positive Features to Look For *(colour plates p.86)*

The physical (type) features to look for are as follows:

1. Body: Very short, thick, compact in appearance
   Great width at the shoulders
   Prominent wing butts
   Body tapering towards the tail
   From the side view, the body should have an egg-shaped appearance, with the top (back) flat
2. Back: Short and flat
3. Wings: Carried close to the body
   Well rounded on the sides and not flat
4. Tail: Medium length
   Slightly drooped
   Sickles and furnishings short and hard
5. Legs and feet: Heavily boned, firm, straight in the toes, yellow in scale colour
6. Shanks: Short
7. Thighs: Extremely stout
8. Head: Strong, wide, powerful in appearance
   Pea comb
9. Eye: Bold and pearl coloured.

The standards call for weights of around 3.6 kg for the male and 2.7 kg for the female in the large fowl. In the bantams, 1.4 kg for the male and 1.1 kg for the female.

Now turning to colour. Remember, the difference between the Cornish (Dark) and the Jubilee is the colour of the lacing. The former is black laced and the latter white laced. The following remarks apply to the ground colour in each variety.

Male

In the Cornish the bird is basically black with a rich sheen and in the Jubilee, white. Some chestnut colouring occurs on the neck, saddle, shoulders and wing bows.

Female

The female ground colour is chestnut, nut brown or mahogany brown. Whichever tone, it is should be even in colour across the bird. The lacing should be sound across the body feathers and double laced for preference. In the Cornish, the head, neck hackle and throat should be green black. The tail needs to be black. On the other hand, the Jubilee is white where the Cornish is black.

The above remarks also apply to the Blue Laced variety which is the blue laced version of the Cornish colour. The under colour is slate.

Negative Features to Avoid

Physical features that need to be avoided or culled out:

1. Poor shape
2. Crooked breast
3. Narrow body
4. Long backs
5. High rounded backs or those with protruding 'kidney' bones ('hippy backs')
6. Bow legged birds
7. Excessively short thighs
8. Flat shins
9. Twisted toes
10. Narrow skulls
11. Long sunken faces
12. Weak looking beaks
13. Grey at the root of the tail
14. Grey in flight feathers.

Looking in the area of colour, weed out those with the following:

1. Males with 'ticking' on the breasts
2. Females with peppering of the ground colour. This is very difficult to breed out once it is found in a strain. It is most obvious in the ground colour between the first and second lace. The same fault also occurs in the Jubilee variety giving the ground colour a 'washy' or 'mealy' appearance.

Breeding Hints

1. There is a deal of debate as to whether Indians should be double mated, but most leading breeders involve themselves with this technique to some degree, so what follows is some idea of how this is undertaken

(a) *Cockerel breeding pen*

The exhibition type male should be mated to a female with dark ground colour, lacking distinct lacing and possessing lots of rich green–black colour.

(b) *Pullet breeding pen*

The female needs to be an exhibition type with particular emphasis on evenness of lacing colour across the body. The male required should show as much female feathering (lacing) as possible with red on the back and in the hackles. Lift back the feathers on the back and look for further evidence of lacing.

2. Jubilee and Cornish are interchangeable in the breeding pen so both varieties can be raised out of the one breeding pen.

3. Most breeders favour introducing new blood on the female side of a strain if fertility appears to be waning.

4. Indians appear to breed better in warmer weather.

5. The breeding pen must be large enough and have sufficient deep litter to allow the birds to effectively exercise or they simply become fat and lazy breeders.

6. Emphasis on overly short legged birds can lead to problems with males being able to effectively tread hens. Some people have got around this by using artificial insemination but one can only wonder at the direction this may lead the breed in general.

The Indian Game breed would have to be regarded as the 'macho' of the poultry world, capturing attention wherever it is exhibited.

Japanese

Breed History

The Japanese bantam is a true bantam with no equivalent in large fowls. It is regarded as an ornamental breed which few people would not consider attractive, hence its long term popularity. Believed to have been a native of the Indo-China region, it was imported into Japan, which accounts for its name. It has the shortest legs of all fowl breeds.

The breed arrived in Great Britain in the early 1800s, seemingly simultaneously from Asia and Europe where it had also been resident for some time. Japanese are recognised in the British Standards in many colours: Black Tailed White, Black Tailed Buff, Buff Columbian, White, Black, Grey, Mottled, Blue, Cuckoo, Red, Tri-coloured, Black Red, Brown Red, Blue Red, Silver and Gold Duckwing, as well as some secondary colours, e.g., Ginger and Furnace, also being permitted. Many argue that the obsession with colour has led to a decline in type as breeders seek to produce new and different colours.

Japanese bantams found their way to the USA at an early date, as the Black Tailed White variety was recognised by the first American Standard published in 1874. The current American Bantam Standard recognises Japanese bantams in sixteen colours.

In Australia the breed has its own specialist club. The date of the breed's arrival in Australia is unknown to the author.

Positive Features to Look For

The following features should be looked for:

1. Body: Short, wide and deep
2. Head: Large, face red
   A gap needs to exist between the tail and the comb
   Male's comb—large, evenly serrated (4–5), coarse textured
3. Tail: Large and fanned
   In the male there should be two long sword-like sickles, slightly forward of perpendicular, many side hangers
4. Saddle: In the male well filled out with hackle
5. Back: Extremely short
   In the female the back line (neck, back, tail) should follow a 'U'
6. Breast: Well developed and rounded
7. Wings: Large, carried drooped not clipped
   Often quoted that the bird 'stands on its wings'
8. Legs: Exceptionally short, stout and free of feathers.

The bird must feel as though it has good stout bone in its body, not thin weak bone, when it is handled.

White Japanese bantam female. Fairfield Poultry Club Autumn Show 1988. Owner: D. Clarke-Bruce.

Black tailed White Japanese bantam male. 1988 Sydney Royal Easter Show. Owner: R.P. Towner.

Negative Features to Avoid

The following need to be avoided or culled out of stock:

1. Tails that sit sidewards rather than perpendicular
2. Wry tail
3. Low tail carriage
4. Tucked up wings
5. Combs leaning to one side
6. Defectively cut combs
7. White patches in lobes
8. Long backs
9. Other than short legs
10. Narrow bodied birds.

Breeding Hints

1. The low stationed Japanese carry a lethal recessive gene such that when they are mated 25% of the offspring fail to hatch, dying in the 18 to 21 day period of incubation. The 25% is, of course, averaged over a large number of chickens hatched. 50% of the chickens will have the desired exhibition leg length and the remaining 25% are regarded as long legged and unsuitable for exhibition.

2. If a long legged male is mated to an exhibition female, 50% will have the desired exhibition leg length whilst the remaining 50% will be long legged.

3. Most colours, it is claimed, breed true but the Black Tailed White often needs a balanced mating approach similar to other Columbian type feather patterns.

4. Since Japanese are short legged birds, they must be kept sheltered and on dry litter. If the litter becomes damp, staining of light underplumage becomes a problem.

Langshan

Breed History

The forerunners of the Langshan breed as we know it today were imported into England and Australia during the mid 1800s from China. These fowls were black, representative of the fowls in their native state where the black colour is regarded as sacred, being a symbol of perfection, majesty, honour and greatness. An example of the importance of this colour in Eastern culture is the black belt of the martial arts. White, on the other hand, is a sign of mourning and if these birds appeared in Chinese flocks they were killed.

The most famous of the European imports from China were those attributed to a Major Croad around 1872. When the progeny of these fowls were first exhibited, judges mistook them for Black Cochins. Much controversy resulted as to what direction the breed should take, resulting in the development of what we now know as the Croad Langshan type and the tighter feathered 'Modern' Langshan.

Some of the Croad style of Langshan were imported into the United States of America where they were accepted as Black Langshans in the American Standard of Perfection in 1883. Some ten years later in 1893, the White variety was admitted to the American Standard of Perfection. Blue Langshans have been recorded as being produced in 1890.

Returning to the English front, one of the reasons for importing the Langshan in the first place was the fact they laid brown eggs. The brown egg laying ability was utilised when clean legged Croads were used in the development of such breeds as the Black Orpington, Australorp, Welsummer, Barnevelder and other European breeds. The White variety of Croad appeared in England around 1886 but has never gained the favour accorded to the Black variety.

The Langshan has an important place in Australian commercial poultry history. The first importations are recorded as having taken place in 1881, and by 1883 they were exhibited at the Poultry Club of N.S.W.'s Annual Show. At the same show in 1896 there were 100 Croads exhibited.

The origin of the Chinese Langshan as we know it can be traced back to importations by a Mr C. Wakfer of Chatswood, N.S.W. in 1905. They appear to have been selected in China along utility lines for their egg laying ability as six pullets entered in the 1906 Hawkesbury College Egg Laying Competition won first prize. This caused enormous interest at the time resulting in a period of careful breeding and selection over the following twenty years, culminating in what we now know as the Chinese Langshan, despite the fact that it was really an Australian creation. In the making of this breed, the Wakfer blood lines were crossed with existing Langshan lines together with some utility Black Orpingtons. The British Poultry Standards do not have a Standard for the Chinese Langshan so it was left to the Langshan Club of Australia to draw up an appropriate Standard which is still used to assess the breed in this country today.

Of the other colours of Chinese Langshan, Whites appeared around the early 1920s, but it is claimed that they never reached the Standard of the Blacks, often lacking in front. The Blue variety was developed independently in Australia by E.J. Winton prior to World War I by crossing a Blue Andalusian hen with a Black Langshan male and mating the resulting progeny back to pure Langshan lines. As with many blue varieties using Andalusian blood, the Blues do not breed true. They follow the inheritance pattern of the Blue Andalusian which exhibits incomplete dominance of colour.

Cuckoo Langshans were reported as being exhibited at the 1946 Sydney Royal Show, catching the eye of the *Poultry* newspaper reporter of the day who commented favourably on their type, but nothing is known of their origins or current status.

Today, in Australia, the Chinese are the most popular variety with a few Croads still kept by enthusiasts and Modern Langshans are regarded as extinct.

As with most popular breeds of large fowls, a bantam version would have to be developed. Interestingly, the first reported Chinese Langshan bantams were developed without using any Chinese Langshan in their make-up. This was achieved in the 1920s by crossing Black Pekins, Black Old English Game bantams and Orpingtons. Another breeder in the 1960s, however, did use Chinese Langshan blood in his creation of bantams by using Black Pekin, Black Old English Game bantams and a small Chinese Langshan cockerel. With such a pedigree, it is easy to see why the occasional red eye appears even in the best line bred stock and why the occasional bad tempered male is bred.

Croad bantams were created in England in the late 1920s and early 1930s by using Black Pekin and Modern Game stock.

Of the bantams, the Black Chinese Langshan is the most popular.

Chinese Langshan

Positive Features to Look For

1. The Chinese Langshan has a distinctive head and particular attention must be paid to it. The comb should be medium sized, straight, of fine texture and rise clear of the back of the skull. The Standard calls for five to six even serrations in the comb. The face should be clean in appearance and free of wrinkles, feathers or hairs. The eye needs to be large in size and dark hazel brown in colour with no overhanging eyebrows. The lobes should be red with no permanent white showing.

2. Turning attention to the body, the back needs to be reasonably broad, flat and of medium length. The front should be well rounded and reasonably deep. Viewed front on the bird should be well rounded from shoulder to shoulder. The keel should be carried level. Medium length wings need to be carried close to the body. The bird should stand squarely on its feet.

3. The shanks and feet should be blue–black in colour with pinkish pigment between the scales. The toe nails need to be white. The soles of the feet white to pink in colour with pink preferred.

4. The feather must be broad.

5. The tail needs to be set at 35° to the body. In the male, the sickles should curve over the main tail feathers.

6. A mature male bird would weigh around 3 kg whilst the females should weigh around 2.5 kg.

Negative Features to Avoid

1. Avoid tall leggy or stout dumpy specimens. Look at the hocks and check for those that are 'in-kneed' or 'cow hocked'.
2. Yellow legs.
3. White beak.
4. Yellow, red or black eyes.
5. Five toes.
6. Permanent white in the ear lobe.
7. Vulture hocks.
8. Blue legs in growers.
9. Wry or squirrel tail.
10. Side sprigs in combs.
11. Crooked breast bones.
12. Feathers on the middle toe.
13. Underweight fowls.
14. Loose or fluffy feathering.
15. In the females, a cushion or fullness in the saddle.

Large Chinese Langshan female. Camden Poultry Club Annual Show 1988. Owner: L.J. King.

Large Chinese Langshan male. Camden Poultry Club Annual 1988. Owner: G. Childs.

Chinese Langshan bantam male. Bantam Club of N.S.W. Annual Show 1988. Owner: G. Frazer.

Black Langshan bantam female. Fairfield Poultry Club Autumn Show 1988. Owner: Stuart Fraser.

Breeding Hints

As with all breeds, type must be the first consideration and when this is under control, attention can be turned to the question of colour.

1. Black

Cull out birds showing blue, purple or brown barring on the feathers. This can occur when persistent breeding of heavily sheened birds is undertaken. This can be overcome by using a 'flat' black bird in the breeding pen.

Any birds showing white feathers should be culled.

2. Blue

The body colour is light blue with dark lacing. The male's hackle, saddle and tail are dark blue.

Blue to Blue matings are not always successful as the blue colour is lost. It appears that better results are obtained by mating dark Blue females to White males and using washy or light blue females to a Black male.

The desired Black male here is one with a dull black with little or no sheen. Blue bred Blacks should not be used in breeding the Black variety.

3. White

This variety is apparently difficult to breed to colour and maintain the dark eye, as it tends to lighten. This corresponds with the natural tendency in white coloured fowls to have a red or light coloured eye. This no doubt would offer a challenge to those who enjoy something difficult to breed, but be sure to discard any brassy coloured birds.

4. Bantams

Because of the breeds used to produce Chinese Langshan bantams, most problems can be traced back to the Black Pekin and Old English Game blood, for example, light eyes, poor type and loose feathering.

Croad Langshan

Croad Langshans have appealed to fanciers over the years for their brown eggs, winter laying persistency and the breed's table properties.

Positive Features to Look For

1. When viewing the male from the side, the tail should be level with the head. Likewise, in the male there should be a balance through the middle of the bird between the front (head and breast) and the rear (body and tail).

2. When viewing the female from the front-on position, the bird should have fullness of breast and width of tail.

3. Croads come in only one colour, Black, and so the plumage should be a dense black with a beetle green sheen. The undercolour needs to be a sound grey.

4. Croads are a large breed with adult males needing to be up around 4 kg and adult females 3.2 kg.

5. A Croad's head should be clean, wrinkle free, with a bright red comb, face, wattles and ear lobes. The beak is light brown in colour. The eye desired is large and brown to very dark hazel, the darker the better.

6. The feet and shanks should be blue-black in colour with pink showing between the scales. Feathers should be present down the shank and along the outer toe of each foot.

Negative Features to Avoid

1. Fowls showing the following faults should be culled or avoided: wry tail, narrow rump, leggyness as youngsters, sloping back when viewed from the side.

2. Cull out the following body colour faults: purple or blue sheen on the black feathers; purple barring; bronze, brown or rust in the body plumage. This latter fault should not be confused with red colouration in the hackle, mentioned for colour breeding. Other body colour faults include: white in body feathers, white or grey in flights and white in the underparts of the neck. The only exception to this is the odd patch of white in the outer toe feathers.

3. Weed out undersized females.

4. Head faults that need to be eliminated include lop comb, side sprigs, permanent white in the ear lobes, yellow around the base of the beak, yellow eye ring and grey eye.

5. Feet and leg faults to be avoided include: yellow legs, yellow or black soled feet, brown or black patches on toe nails which should be white.

Breeding Hints

1. As with many black breeds, a male showing a little

red in the neck hackle is ideal in the breeding pen, not the show pen, to maintain the dense black and desired green sheen.

2. It is important to select large bodied females in the breeding pen if size in the progeny is to be maintained.

3. A couple of important points need to be made with regard to the colour of the chickens and growers. Firstly, do not be put off by the colour of the chickens when they hatch. They appear black with yellow or white down and more often with more white than black. To add to this confusion, legs and toes are covered with yellowish white down and not the expected dark down! Secondly, white can persist in growers, especially in the flights, but these are shed when the adult plumage shows through. Some carry white up until they are 6–7 months of age.

Modern Langshan

A Standard for Modern Langshans does appear in the British Poultry Standards and the Langshan Club of Australia's Handbook, but no Standard appears in the American Poultry Association's Standard of Perfection so we can assume that Moderns do not exist in the USA. The breed is regarded as extinct in Australia.

The Modern Langshan was different from the Chinese and Croad Langshans in the following areas:

1. The Standard weights were heavier with 4.5 kg being called for in the adult male and 3.6 kg in the adult female.
2. The shanks were longer.
3. Compactness of plumage, although this is reflected somewhat in the Chinese Langshan.
4. Low tail carriage in the female.

The breed was bred in three colours, Black, Blue and White.

Moderns have been mentioned as there have been suggestions that some Langshan fanciers have been trying to breed these from long legged Chinese Langshans and they may again appear in the show pen.

Leghorns

Breed History

The White Leghorn has formed the basis of the commercial egg laying industry throughout the world, initially as a pure breed for a long period of time, and of more recent times as the basis of the modern hybrid laying fowl, the White Leghorn–Australorp cross.

Leghorns were produced in many colours including White, Black, Blue, Brown, Blue Red, Buff, Cuckoo, Gold and Silver Duckwing, Exchequer, Mottled, Pile, Red, Partridge and Spangles, but most have now disappeared or are in the hands of Leghorn enthusiasts. Rosecomb Leghorns are to be found in small numbers but Leghorns are best known in their single comb form.

The Americans named the breed Leghorns after the town of Leghorn on the northwest coast of Italy from where they were imported in the early to mid 1800s. The 'Italian Fowl' was imported mainly in the Brown and White forms, but Black, Buff, Pile, Cuckoo and Blue Barred varieties were also imported. At the time, of these importations, the Browns were known as 'Reds'. Leghorns were admitted to the American Standard in 1874 as single comb Browns and Whites.

In the early 1870s the birds were imported into Great Britain from the USA. Later imports were made directly from Italy into Great Britain during the British development of the breed. Whilst the Americans developed a smaller style of Leghorn, the British opted to develop a larger commercial style by infusing White Rock, Malay and Minorca blood into the American bloodlines. The result was a larger Leghorn, and this difference is reflected today in the Standards of both countries. The British Standard specifies 3.4 kg for the cock bird whereas the American Standard specifies 2.73 kg. The hen in the British Standard is 2.5 kg, in the American Standard 2.05 kg.

The development of the White Leghorn in Australia was influenced by the breed's performance in egg laying trials. Interestingly, the world's first twelve month egg laying trials were held at Hawkesbury Agricultural College located west of Sydney. The results of the trials led to the development of the 'Australian' White Leghorn which was a medium sized bird in between the large sized 'English' type and the small imported 'American' type. It was not long before the Australian developed White Leghorn asserted its authority with clear performance advantages over the other two types.

At the beginning of the laying trials, the Browns were favoured and considered the best layers of the Leghorn family. Cuckoo and Buff Leghorns were reported to have performed credibly as well. One interesting import from the USA was the Rosecomb White which performed well but could not match the local product.

As with many breeds, the Leghorn has, over the years, suffered from the extremes of show faddists who have placed undue emphasis on fancy points such as size, combs, wattles and lobes at the expense of the real 'utility' theme behind the breed's standard. This culminated in two classes, 'utility' and 'standard' classes, being set up in shows. This would have been totally unnecessary if breeders and judges had properly applied the standard which had been drawn up with the breed's 'utility' (egg laying) features in mind. Unfortunately, this still persists today.

Accompanying a breed's popularity comes the urge to bantamise it. The Leghorn breed is no exception with most large fowl colours being represented in the bantams. Quite often they find themselves in the winner's circle for major show awards. Backyarders like them because of the egg laying ability and unique character. One drawback is their general trend to non-broodiness, especially if they are bred by the backyarder who would need to keep a broody or two or possess an incubator.

Positive Features to Look For *(colour plates pp.87–9)*

Positive features to look for in a good Leghorn male are as follows:

Overall, the fowl should be well balanced.

1. *Body:* The back should be straight and slightly sloping, the shoulders well developed and square, with the body tapering slightly from the shoulders to the tail. There should be a round full front. The wings carried tightly to the body and ends covered by the saddle hackles. When handling the bird, the body should be flat and straight on the back, breastbone perfectly straight, pelvic bones fine and 2 to 3 centimetres apart.

2. *Head:* The point of the beak needs to be well clear of the front of the comb.
The comb carried erect on a firm, broad base and well clear of the neck hackle. The serrations deep, broad at the base and the largest one over the centre of the skull. Four serrations ensures that they are not too thin ('pencilly'). The wattles should be thin and of a smooth texture. The lobes fairly large, thick, equally matched and of pendant shape. Finally, the eye should be large, red, prominent and fully filling the eye socket.

3. *Tail:* The tail should be large but in balance with the rest of the body. It should be carried at 45° to the backline. The sickles well rounded in form.

4. *Legs:* The thighs should be seen under the bird so it does not have a dumpy appearance. The shanks richly pigmented yellow with close fitting scales. Toes well spread.

Turning to the female, she should be the same as the male apart from the usual sexual differences between males and females. However, attention should be paid to the following. The comb should rise slightly from a firm base and fall over one side without obstructing the eyesight. There should be good length of back,

space between the legs and a low tail carriage closer to the horizontal. The frame should be pliable; breast bone and pelvic bones fine.

Negative Features to Avoid

1. General appearance: Sour or off looking specimens.
2. Carriage: Stiff or stilty carriage.
3. Combs: Folds, wrinkles, thumb marks, side sprigs, double serrations, pencilly (thin) serrations, beefy and rough textured combs. Overlarge combs.
4. Wattles: White spots.
5. Face: White in face, hairy growth on face.
6. Eyes: Beetle or overhanging brows. Sunken eyes.
7. Lobes: Thin or pink tinted lobes.
8. Breast: Flat chested, or cut away fronts. Crooked breast bones.
9. Legs: Feathers on shanks or between toes. Narrowness between legs. Knock knees.
10. Tails: Squirrel tailed males and fan tailed females, especially in the breeding pen as they will produce squirrel tailed males.

Breeding Hints

Much discussion has taken place over the years on Leghorn breeding and it is now accepted that cockerel and pullet breeding pens are required to breed both males and females to exhibition type. Before dealing with each colour, let us consider the type needed for each breeding pen. Firstly, the cockerel breeding pen. An exhibition male is required and also a cockerel breeding female. She has the basic Leghorn features but differs in the following. A small, upright evenly serrated comb, preferably three serrations, carried on

Large Black Leghorn male. Camden Poultry Club
Annual 1988. Owner: L. Simone.

White Leghorn bantam male. Owner: Mrs M.L. Filmer.

White Leghorn large fowl male. Leghorn Club of
Australia Annual Show 1988. Owner: M. Thompson.

White Leghorn large fowl female. Leghorn Club of
Australia Annual Show 1988. Owner: C. White.

Black Leghorn bantam male. Leghorn Club of Australia
Annual Show 1988. Owner: M. O'Connor.

Black Leghorn large fowl female. Leghorn Club of
Australia Annual Show 1988. Owner: G. Sharpe.

Black Leghorn bantam female. 1988 Sydney Royal Easter
Show. Owner: M.A. O'Connor.

a firm base. She should have a well feathered saddle slightly raised to form a 'cushion'. The top two feathers of the tail should extend beyond the others. Secondly, the pullet breeding pen must contain a top exhibition female mated to a male showing the following features. The comb should be weaker at the base, evenly serrated, of fine texture and gently falling to one side. The tail should be deficient in the saddle and show little sickle development. Preferably, the top two sickle feathers just pass the true tail feathers. All other features should be true Leghorn type.

Turning our attention to colour.

1. White
From a colour point of view, only snow white specimens should be bred from. Check plumage carefully for any foreign colour. This however must not be confused with 'sap' which is an excess accumulation of carotene (Vitamin A) from a diet high in green feed or corn. This shows in the shafts of the hackle or flight feathers.

2. Brown
The following is based on the advice given by the late W. (Bill) McGilvray on breeding Brown Leghorns. A cockerel breeding and pullet breeding pen are required to produce the required exhibition colour.

(a) *Cockerel breeding pen*
Male
The male should be an exhibition male with the following colour features.
Neck hackle: Bright rich orange with strong black stripe completely encircled with the orange shade. Head feather not so striped.
Saddle hackle: To match the neck as nearly as possible. Stripe, if present, not as pronounced.
Back, shoulders and wingbow: Crimson red.
Wing coverts: Steel blue with green reflections.
Primaries: Dark brown.
Secondaries: Deep bay on the outer web and black on the inner, the deep bay showing out as the 'wing end' when the wing is closed. Avoid grey in flights.
Breast, belly and underparts: Sound black.

Female
Body colour: Harsh partridge brown with perhaps a lot of rust showing on the wing coverts.
Hackles: Should have a broad black stripe encircled by the golden edge.
Wings: Must be sound in colour.

(b) *Pullet breeding pen*
Female
Body colour: Soft brown partridge, free from 'rust' or 'foxiness' on the wings. Feather must be free

of shaft. Breast colour, salmon red running to a deeper shade about the throat and wattles. Lower breast slightly lighter shade.

Male
Top colour: Washy.
Hackle: More of a lemon shade, stripe much less distinct.
Breast: Patches of brown on the black breast.
Thighs: Brown patches showing on the black.
Wing flights: Light or grey shading should not cause concern.

It is important to keep the strains separated and always line breed from within these strains.

3. Blacks
Arguably one of the most attractive of the Leghorn family, but there are traps for the uninitiated with this variety. Firstly, the youngsters often have a lot of white feathers in their juvenile plumage. Frequently, these can moult out to produce the soundest coloured adults. The lesson here then is not to cull too early on feather colour. Secondly, the leg colour can play tricks. Many stay mottled or green up to 4 months of age and then undergo a change which has them rich yellow by 6 months.

For sound coloured Blacks, a cockerel breeding and pullet breeding pens are required. The cockerel breeding male is the best quality exhibition male available. Pay particular attention to his comb. His mate is a good cockerel breeding Leghorn type, sound in colour but with green or willow legs. Look between the toes and on the soles of the feet as the yellow colour should show through. Avoid females with lead, black or white coloured legs. When talking about white legs, we are not talking about the female who has lost her leg colour pigment following a heavy laying season. It is the naturally white legged type. In the pullet breeding pen, we need an exhibition type female with a strong red coloured eye and bright yellow legs. To this female, mate a pullet breeding male with a red eye, white in the tail especially around the base of the feather, undercolour which is white as snow, deep yellow leg colour and a green top colour, particularly across the wing bar.

4. Blues
The late Jack Salmon was the acknowledged master of this variety and what follows is his thoughts on breeding Blues. Before beginning it is important to know that Andalusian blood was used to derive this variety in Australia, the point being that the breeding plan follows that for incomplete colour dominance. Back to Salmon's advice.

(a) If a pair of Blues are mated, the offspring appears in the following proportions: 50% Blues, 25% Black and 25% Splashed White.
(b) Blue bred Blacks mated to Blue bred Splashed Whites yield 100% Blues.
(c) Blue bred Blacks mated give 100% Blacks.
(d) Blue bred Splashed Whites mated give 100% Splashed Whites.

One mating that was omitted from Salmon's advice was the Blue bred Black mated to a Blue, which would yield 50% Blacks and 50% Blues. This opens the possibility for a breed to produce two colours from the one mating, with an interest then in two varieties in the show situation. A chick remains the same shade of Blue from hatching until maturity so culling for colour can take place early.

Blues need shading or they go brassy coloured.

If the blue colour starts to fade after successive matings, a dark hackled male can be used to rejuvenate the colour.

5. Buffs

A colour that is most attractive but has fallen by the wayside as it is difficult to breed well. When selecting stock for breeding the following are some points to bear in mind. The wings and tails should be free of white. The shade of buff should be even from head to tail, free of ruddiness or brown, cream or mealy tone. The underfluff should harmonise with the top colour. A dark top colour and weak undercolour is undesirable. The quill and the web should be the same colour. Watch the tendency for the neck hackle to be brighter. Tail must be free from bronze or black peppering. Peppered flights or tails in a pale coloured bird are regarded as objectionable.

The following is the order of preference for putting together breeding pens, from most desirable to least desirable. Firstly, male and female both of exhibition shade. Secondly, a dark shaded female to a light shaded male. Thirdly, a light shaded female to a dark shaded male.

When culling Buffs on colour, do not do so until adult plumage is in place, then use the following order to culling with the first named culled first.

1. Peppered or grey flights.
2. Black, grey or heavily peppered tails.
3. Peppering on coverts or back.
4. Those very shafty on the breast or shoulders.
5. Check the undercolour. Check the cockerels around the roots of the neck hackles and roots of tail.

Assuming that some have survived culling, from 4½ months onwards they need to be shaded from the elements. Some old timers used to avoid washing Buffs as well, if they had been kept in spotlessly clean conditions.

6. Piles

There are very few about these days. Those who breed them follow a double mating program similar to that used in Game fowls such as Old English Game as Game breeds were used in their original development. There is evidence that the Pile Leghorn was derived by crossing Browns with white in their flights with White Leghorns. The resulting progeny were crossed with Game fowls and subsequently culled for type and colour.

Little is available on the breeding patterns of the other varieties with the exception of the Exchequer Leghorn which appears to be bred following a balanced mating approach, that is, darker birds to gayer (lighter) birds.

Malay

Breed History

The Malay is of Asian origin with its ancestors being scrub inhabitants characterised by poor flying ability, strong legs and tough skin. These features are carried through into the Malay as we see it today, which lacks feathers and is often bare inside the thighs, centre of the breast and tops of the shoulders. Large specimens have been known to stand 90 cm tall and weigh as much as 5.9 kg.

The breed reached Great Britain around the 1830s and was exhibited at the first poultry show in 1845. Since then it has been used in the background of many breeds on both sides of the Atlantic. In Great Britain, Malays are standardised in Black, Black Red, Pile, Spangle and White. Bantams have also been standardised there but breeders have problems keeping the weights down to those required by the Standard without losing type. In the USA, the Black Breasted Red Malay was admitted to the Standard in 1883 but the others, Spangled, Black, White and Red Pyle, have only recently been admitted in 1981.

The first Malays were thought to have arrived in this country in the 1870s and have had their stalwart supporters over the years. However they could now be regarded as a 'rare and endangered' breed. Malay bantams are not known to exist in this country but there have been a couple of suggestions put as to how they could be produced. It appears nobody has taken up the challenge.

Positive Features to Look For *(colour plates p.90)*

1. Carriage: Very erect, high in front and drooping behind.
2. Body: Very wide at the shoulders, short and tapering towards the tail.
3. Breast: Deep and full, bare of feathers on the breastbone.
4. Neck: Long with a characteristic curve.
5. Shoulders: Carried high and forward, shoulder points very prominent.
6. Back: Curved or roached.
7. Wings: Short and close fitting.
8. Tail: Drooping, narrow sickles and full side hangers.
9. Skull: Very broad, cruel overhanging beetle brows, with a morose expression.
10. Comb: Small, knob shaped, set well forward on the skull.
11. Beak: Curved, powerful, short, hawk-like.
12. Eyes: Set well in the skull, pearl or yellow in colour.
13. Legs: Set well forward to give the inclined body needed.

14. Thighs: Long and powerful.
15. Shanks: Long.
16. Toes: Long and straight.
17. Feather: Very short and hard.
18. Beak and shank colour: Yellow or orange.

The female type is the same as the male but the carriage of the body does not droop so much behind. Tail is short and not fanned.

Negative Features to Avoid

Virtually anything that is opposite to the foregoing needs to be avoided or culled. Three faults that do occur frequently are small heads, straight flat backs and high tails.

Breeding Hints

Heavy emphasis should be placed on type. The breed has a distinct type that must be adhered to or the end result is not a Malay. Colours standardised basically following those of the Old English Game but it must be remembered that only 6 points out of 100 are allocated to colour, emphasising the importance of breeding and judging this breed on type.

The breed has a poor reputation as a layer, and in fact has been nicknamed 'may lay' by some cynics!

The hens make fearsome mothers, staunchly defending the chicks against rats, cats and even dogs. Others are regarded as totally untrustworthy and will kill hatching chickens. It is for this reason that Malay breeders often hatch eggs under broodies or resort to using an incubator, against the grain of many die-hard game fowl fanciers.

If 'macho', rugged Game fowls with a colourful history appeal to you, this could be just the breed to have range on your hobby farm or farm, for this is where they do best.

Minorca

Breed History

The Minorca is the largest of the Mediterranean family of fowls and lays the largest, fine textured eggs of the group. It was also originally known as the Red Faced Spanish before its development in Britain as a utility egg laying breed of fowl. The earliest importations of Minorca stock from Spain were made as early as 1830. These were bred and developed around Cornwall many years before they attracted the attention of Victorian fanciers.

The 1883 Crystal Palace Show signalled interest in the breed when classes were provided and duly subscribed in reasonable numbers. By 1888 a Minorca Breed Club was established and at their first show 144 entries were received for the six classes on offer.

The breed was developed in single and rosecomb form. Many colours were produced including Black, White and Blue as well as the now extinct Cuckoo and Buff varieties. The Black single combed variety is the main survivor today.

It is a pity that, as a result of concentrating on specific physical features, many of the egg laying or productive features of this breed were overlooked, and it was left to the breed's close cousin, the Leghorn, to take up the commercial egg laying stakes. The British Poultry Standards responded by specifying maximum sizes for the fowl's lobes. Evidence of obsession with the breed's lobes can be seen by the contemporary vogue of lobe conditioning treatments which included lime cream, glycerine, zinc cream, petroleum jelly and, if all else failed, saliva! The preoccupation did not stop at the lobes. Excessive comb development was encouraged to such an extent that a cock bird would be unable to hold up his head. He would eventually die because he could not feed or drink effectively! For such fowls to survive they had to be dubbed, a job that had to be done properly if they were not to bleed to death. No wonder the breed fell into disfavour.

Minorcas were imported into Australia in the 1890s, establishing a strong breed following shortly after. Big prices were paid for selected imported specimens with the highest recorded price being £50 for the 1911 Crystal Palace winning pullet. As was the case in Great Britain, the breed suffered at hands of the faddists with the result that today it is in the hands of only a few die-hard enthusiasts.

Bantams were developed prior to World War I by the use of large lobed Black Rosecomb bantams and Black Old English Game bantams. This probably accounts for the problem in Minorca bantams of excessively long wings which are not carried well up against the body as desired by the breed Standard. However, these bantams do figure from time to time in the major awards in Great Britain.

The Minorca is a breed deserving of more enthusiastic and genuine breeder support in this country.

Large Minorca male. Fairfield Poultry Club Autumn Show 1988. Owner: M. O'Rourke.

Large Minorca female. Fairfield Poultry Club Autumn Show 1988. Owner: M. O'Rourke.

Positive Features to Look For

Minorcas should be upright and graceful despite being the largest member of the Mediterranean group of fowls. The male should have a proud strut, although this is difficult to assess in the show pen and is best seen in the yard or free range situation. Minorcas are longer bodied and taller than the Leghorn. They display a more horizontal carriage with the tail rising slightly from the back and, in the male, full and flowing. The feather should be broad and close fitting and possess a healthy sheen.

The white lobes against the black background and bright red face are probably the next most attention getting feature. The lobe is commonly referred to as being almond shaped with the widest part near the top. The British Poultry Standards specifies maximum sizes and they are as follows:

Males: 6.88 cm long and 3.75 cm at the widest part.
Females: 3.75 cm long and 3.13 cm wide.

The lobes should be close fitting, of kid-like texture but with plenty of substance and, of course, pure white in colour. A little vaseline rubbed into the lobes weekly in cold weather will help preserve their condition.

Turning to specific features and starting with the head, the comb is characteristically large but evenly serrated. The male's comb should be firm and upright with the female's gracefully folded over to one side but not obstructing the vision from that eye. The wattles should be long, wide and rounded on their ends. The Minorca eye needs to be full, rounded, bright and dark in colour. The beak is of moderate length, strong and curved. There must be plenty of front with the breast full and rounded. The neck is rather long and arched, carrying plenty of hackle feather especially in the male.

Casting an eye over the back should reveal width at the shoulders, flatness across the topline but tapering towards the tail. The wings are moderately long, carried well up and fitting closely to the sides. The tail, in the female, should be carried well back and be fairly long.

Minorcas lay large white eggs. One Minorca enthusiast claims that the females cackle less when they have laid, thus not annoying sensitive neighbours. If this is the case it would suit many backyarders, but the author has found no hard evidence to substantiate this claim.

Negative Features to Avoid

The British Poultry Standards lists the following as defects in the Minorca breed:

1. White or blue in the face
2. Feathers on the legs
3. Other than four toes
4. Wry or squirrel tail
5. Coloured feathers other than those of the specific variety
6. Incorrect leg colours, that is, they should be
   Black or slate in the Black variety
   Blue or slate in the Blue variety
   White in the White variety.

Other specific problems that do occur with Minorcas include the following:

1. The lobes are difficult to maintain into the second year and as a consequence Minorcas are regarded by many as 'one season' birds. Consequently a

constant stream of youngsters coming along is necessary to maintain a show team.

2. Double folded combs in females
3. Flat shins
4. Large, stilty looking birds often with 'in knees' (cow hocked)
5. Dull colour in the females
6. Whippy tails
7. Over leaning combs in males
8. 'Thumb marked' combs
9. Double serrations
10. Lobe faults to be on the look out for include: white streaks or patches behind the lobes, bluishness in the lobes, hollow lobes (show a concave depression in the middle of the lobe), creases, wrinkles or slackness in lobes.

Breeding Hints

1. With a relatively large breed it is important to use active types and avoid the large drone type. To assist with this, some breeders prefer to use a cockerel and first year hens (female in her second laying season). The hens could then have their performance checked as pullets before committing them to the breeding pen.

2. Size is of great importance in this breed and it is of interest that the hen can weigh as much as the cock bird. Since size is inherited from the female side of the mating a large female is desirable, but coarseness should be avoided. The mature male and female can weigh up to 3.6 kg. The bantam weights are proportionately smaller at 960 g for the male and 850 g for the adult female.

3. A responsible attitude is required to breeding lobes and combs. Attention needs to be paid to the Standard size, thickness and shape. With the comb, especially with the male, a firm comb base is essential if the fowl is to carry a large single comb. Care should be taken not to breed birds with excessive combs as these cause problems, generally in the second year, necessitating dubbing for the fowl's wellbeing. There is no more disgusting spectacle than a cock bird lying in a pen because his head is too heavy to carry around.

Obviously this causes considerable unnecessary stress to the bird. It is an area of breeding that does require common sense and it is the author's opinion that the writers of the Minorca Standard did not have this end result in mind. Judges also need to be alert to their responsibility in the matter and should not favour the exaggerated Minorca.

4. Care should be taken not to put two extremely green sheened birds together in the breeding pen as the offspring will have an undesirable purple sheen instead of the desired lustrous green sheen. A flat or matt black bird is very useful as a breeder in this circumstance.

5. Two separate pens are needed to breed exhibition males and exhibition females.

(a) *Exhibition male breeding pen*
The male needed is a top exhibition male. To this male, use a female with the following features:
(i)   one which stands tall on her legs.
(ii) has a firm comb base carrying a small erect, evenly serrated comb.
(iii) the comb must not be 'overshot', that is, go past the front tip of the beak.
(iv) has a sound face.
(v)   is at the top end of the Standard weight, that is, 3.6 kg.

(b) *Exhibition female breeding pen*
The female needed is a top quality exhibition female with a gently folded comb. To this female, use a male which has large, thick lobes but more rounded, instead of the almond shape desired in the exhibition male. Oversized lobes do not sit on the side of the face but tend to sit partly under the neck and throat. The comb should be clean cut and not too large or the progeny will have too much comb.

6. The newly hatched Minorca chicken's colour may cause some concern to the uninitiated. Black Minorca chickens are white on the throat, breast and underparts. The rest of the body is black. Chicks that hatch black all over never get the desired brilliant black plumage. Some white feathers may appear in the growers' plumage, especially in the flights. These should not cause any great alarm as they will moult out with the adult plumage.

Modern Game

Breed History

These fowls were at the zenith of their popularity 80–100 years ago and illustrate one direction game fowl breeding took after the abolition of cock fighting in England in 1849. Whereas one group of game breeders and enthusiasts sought to perpetuate the ideals of the pit fowl, others turned their attention to developing the 'Modern' game fowl. 'Modern' refers to a style of bird which is long legged, tall and reachy. Many other breeds were also subjected to this development in the Victorian craze for such fowls eg. Modern Langshans.

It is commonly held that Modern Game bantams existed before Old English bantams and in fact were important in the subsequent development of the Old English Game bantam many years later.

A high regard for the breed and its development is evident in the words of the late Modern authority Cecil Thompson, who was quoted as saying that 'the Modern was like the racehorse and greyhound, the aristocrats of the sporting fraternity—lean, hard and fit'.

The Modern existed in both large fowl and bantam but today the large fowl no longer exists in this country. The bantams too are few in number, mostly in the hands of die-hard enthusiasts who regard them as the ultimate development in the poultry fancy. To others they represent a 'freak' with no utilitarian value and probably poultry breeding at its worst. Some claim that the high prices paid for birds in their heyday and the emphasis on exaggerated features spelt the end for Moderns many years ago.

Being an exhibitor's fowl, they need training to perform for the judge to show their desired breed features of tallness and reach. Training requires time and the bird getting used to the judging stick. The usual procedure is to rub the bird up and down the shank, then up to the breast and up the neck to obtain the bird's full reach.

Male birds need radical dubbing before they are exhibited to achieve a 'snaky' head. This point should be considered before taking on the breed. If you have any doubts about dubbing do not take this breed on as the dubbing required is more severe than that required for Old English Game.

Today, Moderns are generally badly inbred and lack stamina, being supported by only a handful of key breeders across the country. Gone are the shoulders, shortness of body, round bone, reach, fitness and hardness that were a hallmark of past specimens.

Positive Features to Look For *(colour plates p.90)*

When looking at a good Modern, the first impression should be of a tall, reachy, balanced fowl with a fiery eye, finely drawn features and fine neat bone. These are the features which stamp the breed as different from all others.

Turning to specifics, traditionalists describe the body in 'flat iron' terms, but this tends to impart a two dimensional image of the desired body. As with other Game breeds, it is better to think in terms of the 'pine cone' model. The body should show little depth of keel and taper from square prominent shoulders. The back should be flat. The wings short, well rounded, fitted to the body and carried up.

The tail is carried at a relatively low angle slightly above the horizontal. It should be tightly whipped and sport fine sickle feathers. The neck, in keeping with the breed, should be long and fine.

The head is lean and snaky, the beak long and slightly curved. To finish the head there is a bold fiery eye.

The thighs should be placed apart to balance the cone shaped body. They should be long and round. The shanks should feature round shins and tight scales. The toes must be long and straight to support this tall fowl. To assist in balance, they must be well spread.

Negative Features to Avoid

1. Overlarge specimens
2. Coarseness of type
3. Coarseness of bone
4. Concentration on length of limb
5. Goose wings, that is, the wings cross over the back of the bird.
6. Lack of reach
7. Faulty limbs.

Breeding Hints

Before looking at colour, which plays a more significant role than in Old English Game, two important points must be remembered. Firstly, Modern females are generally regarded as not being good mothers so young chicks should be raised using a broody such as a Wyandotte × Silkie or artificially reared by themselves and not with other breeds. Secondly, it is important to stick to a strain and avoid outcrossing, unless carefully considered and the outcross follows a similar strain to your own.

Colours used to be well represented with Black Red, Duckwing, Pile, Brown Red, Birchen, Black, Blue and White. Only the three most popular colours of Black Red, Duckwing and Pile will be discussed here.

1. Black Red

This colour is bred in cockerel and pullet breeding lines. Whilst a Wheaten cross, that is, an exhibition male mated to a Wheaten female, is supported by some, many believe that only the Partridge strain should be bred from. If the Partridge line is followed, the exhibition male should be mated to the following female from a Partridge cockerel breeding line. She should be pale, clear capped, clear hackled, a light shade of Partridge with foxy wings. She should have an almost yellowish appearance.

For breeding pullets, use a top quality exhibition female that is perfectly sound in colour, with very fine markings and no rust or shaftiness. The wing ends need to be free of dark bars or lacing. The male needed should have a brickish red colour even in tone from neck to tail with no shading off in the hackles. It is important he comes from a pullet breeding line.

Beware of blue legs in offspring as this is an indication of Duckwing blood in the strain.

2. Duckwing

Double mating is required. Two cockerel breeding matings are as follows. Use an exhibition male mated to a Duckwing female with clear neck hackles, a shade rusty in the wings and a darker but evenly coloured body. An alternative is to use a Black Red male mated to a Duckwing female. The resulting pullets can be mated to a Duckwing male for cockerel breeding but must not be used to breed Black Reds.

When breeding pullets, it has been advocated that a Silver Duckwing male from a pure Duckwing line can be mated to a sound coloured Black Red female. Alternatively, a medium shaded Duckwing male can be mated to exhibition females.

3. Pile

Double mating is needed as with Pile in other breeds. For the cockerel breeding pen, an exhibition male with bright top colour, sound dark chestnut wing bays, pure white breast and wing bars is mated to a sound deep salmon breasted female with red on the wings. Another cockerel breeding pen is an exhibition Black Red male mated to a pale breasted Pile female. The drawback of this mating is that the pullets end up mostly willow legged.

The standard pullet breeding pen is a dark topped male with a marble marked or peppered breast mated to a female clear in the wing and showing a good salmon breast. Cecil Thompson the past master of this breed also suggests the following mating to breed pullets. A Pile male bred from exhibition females mated to perfectly coloured Black Red females whose wings and body should have the same tone.

Whilst colour plays an important role in Moderns, the breeder cannot escape placing top priority on type as without type one does not have a breed.

New Hampshire Red

Breed History

This breed was developed by commercial poultry farmers in the State of New Hampshire in the USA over a period of about thirty years from Rhode Island Red stock imported from the neighbouring State of New York. The development program started in 1915 and concluded with the breed's admission to the American Standards in 1935. The breed was developed on commercial rather than exhibition lines with the following criteria in mind:

1. early maturity
2. rapid growth
3. rapid and complete feathering
4. low chick mortality
5. high hatchability
6. early large egg size
7. high average egg production
8. high vitality
9. low adult mortality.

Part of the program of development was the setting up of pullorum free flocks.

A bantam version is also common in the USA but is not recognised in the British Standards.

In Australia, the New Hampshire Red was introduced by Mr F. Cook of Liverpool who imported eggs before importations were prohibited. Sufficient stock was subsequently produced that some years later they competed quite successfully in the Hawkesbury Laying Trials. Because of the small amount of parent stock that was imported, a sufficiently large genetic pool could not be established to permit commercial exploitation of the bird along the lines of the Leghorn, Australorp and Rhode Island Red industries. As a result the breed has not been as numerically strong as the others. Today there are relatively few New Hampshire Reds in Australia being bred for exhibition.

Positive Features to Look For

The physical type to look for is as follows:

1. Body: Very blocky
   Medium length and not as long as the Rhode Island Red
   A relatively broad but deep body well sprung at the ribs
   The width of the body to be carried well back
2. Breast: Deep, full and well rounded
3. Keel: Long and extending well forward to produce the rounded breast
4. Head: Rugged in character but not coarse
   Refined in appearance and carried erect
5. Tail: The male's tail carried at 45° to the back
   The female's is carried a little lower at about 35°
6. Weights: This is a heavy breed so the desired weight for the male is up to 3.85 kg and 2.95 kg for the female.

Negative Features to Avoid

The breed Standard does not list any faults to be on the look out for, so it can only be assumed that the normal faults opposite to the features in the desired type should be culled out. For example, rounded backs, undersized fowls, shallow bodies, in knees or 'cow hocked' birds and so on.

Likewise any variation to the desired colour, for example, white feathers in the outer plumage, a washed-out buff colour, deep chocolate or muddy coloured birds should be culled.

Breeding Hints

Very little is documented on the breeding of New Hampshires other than that they breed reasonably true and require the normal culling of those that fall short in type or have obvious physical deformities.

This is, perhaps, a breed that could be well suited to the 'hobby farming' belt where a productive breed needs to be kept and occasionally exhibited.

Blue Andalusian bantam male. Bantam Club of N.S.W.
Annual Show 1988. Owner: K. Cook.

Andalusian bantam female. Camden Poultry Club
Annual 1988. Owners: R. & J. Perkinson.

Blue Andalusian bantam female. Bantam Club of N.S.W.
Annual Show 1988. Owner: K. Cook.

Andalusian bantam male. Camden Poultry Club
Annual 1988. Owners: R. & J. Perkinson.

Australian Game Pile large fowl female. 1988 Sydney Royal Easter Show. Owner: J. Rogers.

Black Red Australian Game male. Fairfield Poultry Club Autumn Show 1988. Owner: Mel Pearson.

Duckwing Australian Game female. Fairfield Poultry Club Autumn Show 1988. Owner: Mel Pearson.

Legbar male.

Legbar females.

Barnevelder bantam female. Camden Poultry Club Annual 1988. Owner: G. Childs.

Large Barnevelder female. Camden Poultry Club Annual 1988. Owner: G. Childs.

Large Barnevelder male. Camden Poultry Club Annual 1988. Owner: G. Childs.

Barnevelder bantam male. Camden Poultry Club Annual 1988. Owner: G. Childs.

Product of a breeding program by G. Childs to develop Brahma bantams. Brahma bantam female. Bantam Club of N.S.W. Annual Show 1988. Owner: G. Childs.

Product of a breeding program by G. Childs to develop Brahma bantams. Brahma bantam male. Bantam Club of N.S.W. Annual Show 1988. Owner: G. Childs.

Millefleur d'Uccle Belgian bantam female. Bantam Club of N.S.W. Annual Show 1988. Owner: B. Dukes.

Millefleur d'Uccle Belgian bantam male. Camden Poultry Club Annual 1988. Owner: A. Cheetham.

Lavender d'Anvers Belgian bantam female. Bantam Club of N.S.W. Annual Show 1988. Owner: B. Dukes.

Lavender Belgian Barbu d'Anvers male. 1988 Sydney Royal Easter Show. Owner: Frank Catt.

Quail d'Anvers Belgian bantam male. Camden Poultry Club Annual 1988. Owner: A. Cheetham.

Quail d'Anvers Belgian female. Camden Poultry Club Annual 1988. Owner: A. Cheetham.

Silver Grey Dorking large fowl female. 1988 Sydney Royal Easter Show. Owner: D.H. White.

Large Dark Indian Game female. Camden Poultry Club Annual 1988. Owner: B. Usope.

Large Dark Indian Game male. Camden Poultry Club Annual 1988. Owner: A. Cooksley.

Jubilee Indian Game large fowl. 1988 Sydney Royal Easter Show. Owner: B.G. Weiss.

Jubilee Indian Game bantam. 1988 Sydney Royal Easter Show. Owner: Mrs E.B. Green.

Blue Leghorn large fowl female. Leghorn Club of Australia Annual Show 1988. Owner: G. Sharpe.

Best Blue Leghorn bantam. 1988 Sydney Royal Easter Show. Owners: S. & A. Nicholas.

Cuckoo bantam Leghorn male. Camden Poultry Club Annual 1988. Owner: K. Heydon.

Large cuckoo Leghorn male. Camden Poultry Club Annual 1988. Owner: L. Simone.

Brown Leghorn large fowl female. Leghorn Club of
Australia Annual Show 1988. Owner: G. Lee.

Brown Leghorn large fowl female. Leghorn Club of
Australia's 1987 Annual Show, Newcastle, N.S.W. Owner:
George Lee.

Cockerel breeding brown Leghorn large fowl female.
Leghorn Club of Australia Annual Show. Owner:
L.T. Ford.

Brown Leghorn bantam female. Leghorn Club of
Australia Annual Show 1988. Owner: Ken Bergin.

Brown Leghorn bantam male. Bantam Club of N.S.W.
Annual Show 1988. Owner: K. Bergin.

Newly hatched Black Leghorn bantam chicken.

Blue red Leghorn bantam chicken.

Brown Leghorn partridge coloured chicken.

Large white Leghorn female. Camden Poultry Club Annual 1988. Owner: B. Boardman.

Buff Leghorn large fowl female. Leghorn Club of Australia Annual Show. Owner: M. Lye.

Buff Leghorn bantam male. 1988 Sydney Royal Easter Show. Owner: N.J. Schultz.

Blue Red Leghorn bantam female. B.C. N.S.W. 1988. Owner: K. Bergin.

Duckwing Malay Game female. Fairfield
Poultry Club Autumn Show 1988.
Owner: Noel Middlebrook.

Wheaten Malay Game female. Fairfield
Poultry Club Autumn Show 1988.
Owner: Noel Middlebrook.

Black Red Malay Game male. Fairfield
Poultry Club Autumn Show 1988.
Owner: Noel Middlebrook.

Modern Game Pile bantam male. Bantam Club of N.S.W.
Annual Show 1988. Owner: C. McKenzie.

Modern Game Partridge female. Bantam Club of N.S.W.
Annual Show 1988. Owner: C. McKenzie.

Old English Game black tailed Wheaten female. Bantam Club of N.S.W. Annual Show 1988. Owner: S. Adams.

Old English Game Black Red light leg bantam male.

Large Black Red dark leg Old English Game male. Camden Poultry Club Annual 1988. Owner: B. Ford. Tail has not fully grown out following the autumn moult.

Old English Game Blue Red bantam male. Bantam Club of N.S.W. Annual Show 1988. Owner: R. Guy.

Large Old English Game Pile female. Camden Poultry Club Annual 1988. Owner: R. Cupitt.

Old English Game black Red bantam male. Bantam Club of N.S.W. Annual Show 1988. Owner: S. Adams.

Wheaten Pekin bantam female. Best Wheaten or Black Red, Pekin Club of N.S.W. Annual Show 1988. Owners: A. & S. Mills.

Partridge Pekin bantam male. Pekin Club of N.S.W. Annual Show 1988. Owner: W.K. Dickson.

Brown Red Pekin bantam female. Best Brown Red, Pekin Club of N.S.W. Annual 1988. Owner: F. Fogarty.

Blue Pekin bantam female. Best Blue, Pekin Club of N.S.W. Annual Show 1988. Owners: A. & S. Mills.

Buff Pekin bantam female. Best Buff, Pekin Club of N.S.W. Annual Show 1988. Owners: K. & G. Lambert.

Buff Pekin female. Bantam Club of N.S.W. Annual Show 1988. Owner: G. Wood.

Rhode Island Red bantam female. 1988 Sydney Royal
Easter Show. Owners: B. & E. Bell.

Rosecomb Rhode Island White bantam female. Bantam
Club of N.S.W. Annual Show 1988. Owner: J. Clark.

Large Rhode Island Red male. Camden Poultry Club
Annual 1988. Owner: J. List.

Large Rhode Island Red female. Camden Poultry Club
Annual 1988. Owner: J. List.

Bearded Blue Silkie female. Fairfield Poultry Club
Autumn Show 1988. Owners: S. & D. Lindsay.

Gold Sebright female. 1988 Sydney Royal Easter Show.
Owner: E.J. Flarrety.

Large Welsummer male. Camden Poultry Club Annual
1988. Owner: G. Childs.

Large Welsummer female. Camden Poultry Club Annual
1988. Owner: G. Childs.

Partridge Wyandotte bantam female. Bantam Club of N.S.W. Annual Show 1988. Owner: R. McCredie.

Buff Wyandotte bantam female. Bantam Club of N.S.W. Annual Show 1988. Owner: R. McCredie.

Buff Wyandotte bantam male. Bantam Club of N.S.W. Annual Show 1988. Owner: R. McCredie.

Faults *(black and white photographs on pp.134–6)*

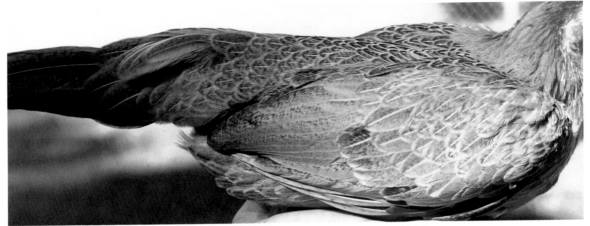

Faults: 1. "Foxiness" or red in the wing of a partridge coloured bird.
2. "Shaftiness" or light coloured feather shaft.

Fault: Red in white lobe.

Fault: Neck hackle does not have the clear distinct white border around the edge of the feather. Breed: Light Sussex.

Old English Game

Breed History

The Old English Game large fowl can be traced back many centuries in Britain. Some evidence points to game fowls being kept by the early Britons at the time of the Roman invasion in 55 BC. Game fowl keeping has been supported by Royalty through the ages, with documented support of cockfighting going back to the reign of Henry II (1154–1189); it was fully recognised as a sport in the reign of Charles II (1649–1685).

An early attempt by Oliver Cromwell (1653–1658) to prohibit cock fighting was duly repealed. The activity subsequently continued to receive support from Royalty and the well to do until it was legally stopped in Britain by an Act of Parliament in 1849.

Just prior to this time saw the start of poultry exhibitions and the Old English Game fowls soon found their way into these activities. The most well supported exhibition was that held at Oxford. This led to the formation of the Oxford Game Fowl Club which was, in turn, responsible for drawing up the Oxford Standard aimed at maintaining the true Game fowl in its original form. A revision of this original standard was made in 1920 and this forms the basis of what appears in the British Poultry Standards today.

With European settlement the breed found its way to Australia, with cockfighting activities being reported in the *Sydney Gazette* in 1810, 1826 and 1830. After its abolition in 1850, cockfighting persisted illegally for the next 100 years, often being patronised by professionals, businessmen and the well to do. Little wonder the breed had such wide support. Today, it is mainly in the hands of Old English Game enthusiasts.

The Old English Game bantam's history is one of controversy. Some have been bred down from large Old English Game, but the most convincing evidence points to them having developed independently as a purely exhibition fowl whilst following the Oxford Standard as closely as possible, that is, a miniature of the large fowl. Today, Old English Game bantam classes are probably the most widely supported of all classes at most shows. The reasons for this popularity could be found in the following:

1. A large range of often striking colours.
2. Often several colours can be produced from one mating.
3. Their alert and cheeky character.
4. Ease of breeding.
5. The hens are good mothers, making the raising of chicks an easy task.
6. They take up little space.
7. They are a hardy breed.
8. Relatively easy to prepare for the show pen.
9. Often figure in major show awards.

Positive Features to Look For *(colour plates p.91)*

Here we shall consider the physical features that make up the shape or type of the breed. The approach to type will be the same for bantams as for large fowls as they 'should be a true miniature of the national fighting Game'. Reference also has to be made to the two schools of thought regarding type that have emerged over recent times. To complicate matters, both schools of thought claim to follow the ideas of the Oxford Standard but the end products of their breeding are two distinctly different fowls which can be readily identified when seen in the one class at a show. One school of thought supports a type closely following that accepted by the old 'cockers' of yesteryear. This bird is agile in action, active as well as having a marvellous flow of feather in the neck, saddle and tail. Much attention is paid to the bird's action and handling. The other school of thought has produced a bird which is wider, with particular emphasis on width across the shoulders, strong heads and size. These birds do not as a general rule possess the flow of feather or agility of the first mentioned school of thought. When deciding to breed Old English Game it would be a good idea to decide which school of thought you wish to subscribe to. This is very important when exhibiting these fowls as there are variations in how the Standard is interpreted and subsequently awards given. Variations of interpretation can often explain why a bird can win in a well supported class one week but be unplaced in an equally strongly supported class the next. This situation has been observed in several cases where almost the same line-up of fowls are judged by different judges in successive weeks! This situation has demoralised many a fancier both new and experienced. The moral of this is to read the Standard closely and breed to it, but also remember that others do not always see it the way you do.

Turning our attention to what we should be looking for. Volumes have been written on type in Old English Game but this is an attempt to describe the required type in a short, concise, meaningful statement.

1. General body shape or type

The most frequently quoted shape is that of the old-fashioned 'flat iron'. However, this leaves the reader with a two dimensional picture rather than the desired three dimensional picture that is necessary to truly envisage the desired shape or type. A pine cone would give a far better three dimensional picture of the distinctive shape and would also help with comprehending the handling features which play such an important part in assessing Old English Game. The 'pine cone model' makes it easy to envisage the desired rounded sides and width at the shoulders. It also gives some idea of the proportions the body should have.

Old English Game Spangle bantam female. Fairfield Poultry Club Autumn Show 1988. Owner: Darcy Roth.

Old English Game Spangle bantam male. Bantam Club of N.S.W. Annual Show Annual 1988. Owner: D. Roth.

Old English Game Spangle female. 1988 Sydney Royal Easter Show. Owner: B.H. Burnham.

Sideways view

he wings should be large and strong, carried closely
o the body, following the line of the cone form. They
eed to be large and strong to protect the bird's thighs
n the sparring attitude.

The back is flat with the neck and tail covering the
ounded ends of the body 'cone' shape.

The thighs should be positioned to balance the body
one' with the hocks angled as if posed for action
vithout giving the bird a crouched appearance. The
rouched position is 'body language' for submission
nd should not be evident in this breed which is noted
or its courage.

The head, an essential part of the bird's fighting
quipment, should be relatively small with a strong
eak capable of shutting tightly together. A bold
xpressive eye is an important feature and should
npart the spirit of the breed. The skin of the face and
eck should be supple but not baggy or loose.

The neck should be of medium length and balanced
o the rest of the bird. It should be stronger at the base
nd well covered with neck hackle feathers particularly
vhere it joins the shoulders.

The tail should be set at around 45° to the back-
ne, fan shaped but more accurately like a partly closed
an, and well furnished, that is with well developed
ickles, side hangers and saddle hackles.

. *The front-on view*

The 'cone' body shape should once again be immed-
ately evident. The legs and thighs are positioned to
alance the cone shape, but not too far apart so as to
ive the bird a clumsy action when it moves, nor too
lose together so that the bird looks as though it would
all over with the slightest push from the side.

The thighs and legs must not be too long or too
hort. Once again, the theme of balance of propor-
ion is important.

The toes should be straight, well spread and strong
o balance the bird's body. It is particularly important
hat the rear toe follow the line of the central toe in
ront.

4. *Handling*

The three dimensional cone model should be evident
when the bird is placed in the hand(s). The bird should
eel balanced. The wings and thighs should contract
strongly back to the body when extended.

Body plumage should be resilient, lustrous and
closely following the contour of the body. It should
indicate the health of the bird.

Shanks should be clean, tightly scaled and sport a
pair of hard, firmly attached spurs.

The body should feel hard and in good athletically
fit condition, not fat and flabby. It could be described
as 'lean and mean'.

In the pen the bird should be alert and active and
move easily and quickly on its feet, as if ready for
action should this become necessary.

The female, apart from natural sexual differences,
should display the same features.

Negative Features to Avoid

The Oxford Standard lists the following as serious
defects to avoid:

1. Neck: Thin necks
2. Body: Flat sides
   Deep or long keel
   Pointed, crooked or indented breast bone
3. Thighs and legs: Thin thighs
   Straight or stork legs
   In knees ('cow hocked')
4. Feet: Thick toes or insteps
   Duck feet (rear toe not pointing backwards in line
   with central toe)
5. Handling: Soft flesh
   Soft, broken or rotten plumage
6. Deportment: Bad carriage or action (movement)
7. Any indication of constitutional weakness

Any bird which is lacking in the points discussed under
'type' should be considered as a candidate for culling.

Breeding Hints

Colour faults will be mentioned as they become
relevant when discussing breeding each of the colours.
It should be remembered that, particulary in Old
English Game, type is of paramount importance and
must be regarded as the first consideration when
selecting fowls for breeding, particularly the female.

The same colour breeding principles apply equally
to large fowl and bantams.

While there are upwards of thirty different colours
in Old English Game fowls only a selection of the more
popular ones will be discussed here.

1. Black Red

These appear in two different strains, the Partridge
bred and Wheaten bred strains. The two strains must
be kept separate and should not be interbred.

Partridge Bred Black Red

These come in light and dark legged strains but the
dark legged strain is most favoured. The following is
a description of the male and female of this variety
that should be mated together.

Male

Hackle: Orange–red, free from striping

Back and shoulders: Dark red

Wings: Deep red on top
    Dark blue bar across

Secondaries: Bay

Primaries: Black ends

Body: Lustrous green–black.

Female

Hackle: Golden streaked with black
    The gold to extend around the striping

Back and wings: Partridge, which must not be bold
    or large in pencilling but should be a small, wavy,
    irregular marking, really a series of small black
    peppered dots on the partridge ground colour

Breast: A rich salmon, running off to an ashy colour
    on the thighs

Tail: Black shaded with brown.

Some breeders will set up separate cockerel breeding and pullet breeding pens with this variety, particularly to try to gain the soft brown partridge colour in the female. The male for such a mating would probably be broken in the breast, that is, have brown lacing showing through and more yellow-orange than orange-red in the hackle. The top colour would be a lighter shade. Likewise, if a cockerel breeding pen was to be set up, the female would be of a darker partridge shade and have 'rust' or 'foxiness' showing in the wings.

Wheaten Bred Black Red

The Wheaten bred Black Red strain only comes with light legs although the odd dark legged bird appears in these lines, probably because of some prior cross with the Partridge bred line.

Male

Hackle: Light golden red, free from striping.

Back and shoulders: Bright red
    The rest of the fowl is the same as for the Partridge
    bred Black Red.

Female

Hackle: Golden red free of black streaks

Breast and thighs: Light wheaten colour

Tail and primaries: Almost black but stipled with dark
    wheaten colour.

This pen is favoured in Black Red classes these days as exhibition males and females can be bred from this mating. Some breeders have been known to further subdivide these into cockerel breeding and pullet breeding lines. The pullet breeding male will have a broken breast with reddish brown lacing present. The cockerel breeding female will carry a darker reddish wheaten colour across her back and wings.

2. Spangles

It is widely accepted that both cockerel and pullet breeding pens are required to successfully breed Spangles. The cockerel breeding pen would consist of an exhibition quality male mated to an over-spangled or 'gay' female. On the other hand, the pullet breeding pen would be made up of an exhibition standard female with fine spangling, mated to a very dark but finely spangled male of good type. Look for a double wing bar of spangling in the male.

The spangle pattern holds a few traps for the uninitiated and impatient as the chicken feathers give no indication of the parental plumage. The spangling changes several times as the fowl grows, so it does not pay to cull on spangling until the fowls are fully grown. To complicate matters, spangling changes as the fowl gets olders, generally becoming gayer as the bird ages. A bird dark in its first year could moult to the correct colour in its second year. With such changes taking place the breeding of top Spangles can be a real challenge.

3. Duckwing

These are found in Silver, Gold and sometimes an odd Blue Duckwing. Once again, double mating is necessary to produce top line exhibition fowls.

The pullet breeding pen requires a top exhibition female and the following male. He should be a dull, soft steel grey coloured bird, rather black across the shoulders, striped in the hackle, free of shaft or rust, have a sound wing bay, a white tipped breast and no white in the flight feathers.

For cockerel breeding, a top exhibition male should be mated to a hen with the neck hackle as clean as possible (free of striping) and heavily pencilled in the body. A dark body colour is useful too.

Two other matings which can produce some interesting results: a good Golden male can sometimes be bred from a very dark female free of rust or from light silvery females that are extra dark in breast colour and very foxy on the wing.

4. Pile

Piles can be successfully bred from Pile × Pile matings but the resulting cockerels lose their colour after several generations, necessitating crossing with Black Red blood to improve the colour once again.

Those who use a double mating program usually proceed as follows. A pale breasted, near white Pile female is mated to a Black Red male to produce brightly coloured cockerels. Top quality pullets can be bred by crossing an exhibition female with a light or lemon Pile male. The resulting females have a clear white body, salmon breasts and gold and white neck hackle. The cockerels produced lack top colour for exhibition.

If the lightest coloured Piles are mated over several generations, Whites can eventually be produced.

5. Blue Red

Blue Reds are similar to Black Reds except that where they are black, Blue Reds are blue (slaty grey). These breed reasonably true. They can also be produced by several crosses. Firstly, a Black Red male is crossed to a Blue Tailed Wheaten (the blue appears instead of the black in a normal Wheaten female). Interestingly, Blue Tail Wheatens can also be produced, as can Blue Red males, by crossing a Wheaten bred Black Red male with a White Tail Wheaten female. Secondly, if blue bodied or partridge marked hens are used the offspring become darker each generation, eventually resulting in self coloured dun blues. Thirdly, a cross to a Wheaten bred Black Red will brighten up the top colour of the Blue Red cockerels if the colour fades with several generations of breeding.

6. Crele

These breed quite true for a while then need 'freshening up' and this can be achieved by the following. To improve the top colour, cross a Golden Crele with a dark harsh coloured Partridge female, but she must be Partridge bred. To improve female colour, a Golden Crele male can be mated to a well coloured Duckwing female. Or to improve the colour of both, mate a Golden Crele male to a Partridge bred Blue Red female.

7. Birchen

Many claim the Birchen has lost its gipsy coloured face over the years and this could be retained or improved by crossing with a Brown Red. It is also claimed that this would retain the jet black eye as well as the hardness and tightness of feather.

8. Black

When selecting breeding pens, look for the brilliant sheen in the feather, gipsy coloured faces (mulberry), dark eyes, beak, legs and feet. Blacks breed true but tend to go 'soft' in the feather. Constant sheen to sheen matings can end up producing 'purpling', so the use of a matt black bird will become necessary.

9. White

As mentioned previously, whites can be produced from successive very light Pile matings. Whites, as with Blacks, tend to go 'soft' in the feather but will breed true to colour.

10. Other colour matings including those that produce the 'off' colours

Some of these matings will produce more than one colour but have been mentioned to highlight the colour attraction of this breed.

A Partridge bred Black Red male mated to a Duckwing hen will produce Golden Duckwing males and Partridge females.

The above male mated to a Blue Red female will produce Blue Red males and an occasional Blue Red pullet.

The same male mated to a Black hen will produce some dark winged Black Red males.

Using a Wheaten bred Black Red male, some interesting results can be produced from mating with a Blue Pile (Blue from a Blue Tail Wheaten) to produce Blue Red males, or with a Pile to produce Pile cockerels.

If a Duckwing male is mated to a Partridge female Black Red males and Duckwing females will be produced. Mating him to a Blue Red hen will result in the odd Blue Red cockerel and occasional blue breasted Duckwing pullet.

The above crosses are some suggested by the late Joe Maude father of the legendary and late Harry Maude.

While the Old English Game breed provides an amazing array of colours, many not mentioned here, attention should be paid at all times to *type*. If good type can be established, it can be seen that with careful mating it is theoretically possible to reproduce this type in many colours.

Orpington

Breed History

The Orpington breed was developed by William Cook in Britain and first came to the attention of the poultry world in 1886. It was developed in an era when the Dorking, Game and Cochin reigned supreme in the poultry yards of Britain. Many traditional poultry people frowned on the new breed which was developed by cross breeding and selecting along utilitarian lines.

Cook's objectives were to develop a breed which laid large clutches of brown eggs with good winter persistency as well as produce a carcass with plenty of white meat. The English have always expressed a preference for white skin and meat in contrast to the Americans who developed the yellow skinned breeds with a tendency to darker meat. In addition to the utility properties, Cook was keen to foster a 'handsome appearance' in his fledgling breed. The Orpington appeared at the world's first laying trial held in Britain in 1887.

Whilst the Langshan fowl was embroiled in much controversy over the direction the breed should take, the Orpington fowl became the victim of 'show faddists' who distorted Cook's original utilitarian objectives to produce a massive 'feather duster' which bore no relation to Cook's original intentions.

It took Australian breeders to identify the utility features of the breed and return it to something like the original lines desired by Cook. The newly selected Australian birds became known as Australorps and were imported back into Britain in the 1920s. This is why Australorps are known in Britain as the 'boomerang' breed; they came back probably more as Cook had originally intended his Orpingtons to be. Cook's influence on the development of the Australorp should not be idly dismissed. In the meantime, the Orpington in its country of origin was developed into a massive bird with extremely loose feathering and devoid of the utility features that Cook had developed in the breed.

Bantams of this breed were developed pre World War I but were mainly Buffs developed on Oriental blood. Black bantams appeared in numbers in the 1930s.

Orpingtons come in four main colours, Black, Buff, Blue and White. Black and Buff today are the most popular numerically in this country. Jubilee, Spangle and Cuckoo varieties were developed but are now regarded as extinct.

The Buff Orpington has the privilege of Royal patronage in Britain where the Queen Mother has had a long interest in the breed and has been a successful exhibitor at important British shows. The Queen Mother also has a trophy for the Best Buff at the National Show to encourage the variety.

William Cook developed the Buff variety by using a local breed, the Lincolnshire Buff to 'clean up the Cochin' influence in the Orpington, that is, loose feathering, feathered legs and shanks. This is probably why the Buffs tend to be slightly tighter in the feather than the other varieties. The Buff became one of Britain's most popular breeds up to 1914 and has steadily declined since then, possibly because its plumage fades easily in the sun and rain, necessitating shading to maintain the Buff colour.

Controversy surrounds the origin of the White variety. Former leading British poultry expert W. Powell-Owen suggests two possible avenues of origin. Firstly, a White Leghorn male crossed with a Black Hamburgh and the progeny crossed with a rosecombed White Dorking, then this progeny crossed with a single combed Dorking. Secondly, a light Sussex with the black feathering bred out.

It may be of interest that moves are underway in Australia to develop the White Orpington using White Sussex blood.

The Orpington in Australia has a breed club based in N.S.W. to support fanciers.

Positive Features to Look For

1. The first impression of this breed should be one of size. When viewed from the side, the overall shape should follow that of a 'U' outline through the curved neck, short back and tail. The tail should be carried high in a rising sweep. The wings should be carried up and not droop. The body needs to show pronounced depth with a deep round front. The head and comb are neat and small. The hocks are entirely covered by body fluff, but not the shanks.

2. When viewed front on, width of body should be obvious with a wide chest, back and saddle. The tail should be full, compact and carried high.

3. The plumage should be fairly close and not too loose. The feather itself must be broad, soft and abundant or profuse. Profuse feathering refers to the density of the feathers rather than the 'looseness' of them.

4. An important feature of the Orpington fowl is its handling. Not only should the bird appear large but it must handle large. The bird must feel bulky and not just a 'bag of feathers'. Upon handling, a large deep keel must be evident. The chest must be broad and carrying plenty of breast meat as Cook, the breed's originator, intended it.

5. The bantams should be true miniatures of the large fowl showing a broad deep and cobby body with the wings carried well up.

6. From a colour point of view, the following should be evident.

(a) *Blacks*
A rich beetle green sheen with the black carried down the feather to the skin.

(b) *Buffs*
A clean *even* buff across the body with the buff colour also carried down the feather to the skin. In Buffs, evenness of colour is the key feature.

(c) *Whites*
A pure snow white colour is required.

(d) *Blue*
The Blue variety follows the Andalusian pattern for blue colouration. The female should be a mid slate blue with a darker shade of blue lacing. The neck hackle should be dark blue. The male needs to have a dark blue top line but the breast and underparts are as for the female.

Negative Features to Avoid

1. It is interesting to note that the British Poultry Standards mentions faking and trimming as disqualifications in this breed.

2. The British Standards further regards the following as serious defects and they need to be avoided or culled out:

(a) Side sprigs on combs
(b) White in the red ear lobes
(c) Feather or fluff on the shanks or feet
(d) Long legs
(e) Any physical deformity
(f) Yellow skin or yellow on shanks or feet
(g) Yellow or sappiness in the White variety
(h) Coarseness in the head, legs or feathers of the Buff.

3. Breed faults could be generally summed up as those opposite the desired breed features and would include:

(a) Ultra low stationed fowls
(b) Excessively loose or fluffy fowls
(c) Short feathered birds
(d) Ball, drooped or narrow cushions in females
(e) Long flowing tails
(f) 'Beetle browed' heads
(g) Downward pointing wings

Large Black Orpington hen. Camden Poultry Club Annual 1988. Owner: B. Boardman.

Large Black Orpington male. Camden Poultry Club Annual 1988. Owner: B. Boardman.

Pair large Black Orpingtons. Camden Poultry Club Annual 1988. Owner: B. Boardman.

Black Orpington large fowl female. 1988 Sydney Royal Easter Show. Owners: B.W. & P. Boardman.

(h) Lanky feathering especially around the thighs and stern.

In general, any feature that makes the bird look shallow, narrow or long.

4. Specific features to be looked for in the varieties include the following:

(a) *Blacks*
Dull black, bronze and purple patches or barring, especially on wings and tails.

(b) *Buffs*
Colour tone differences between hackles and tails. Black or cinnamon in tails or wings. Light undercolour or white flecks in plumage. White patches in wings or tails. Colour fading away at the thighs.

(c) *Whites*
Yellow or straw tinges or 'sap' in the feathers or shafts (quills). The shafts is where this is more likely to occur. Blue or yellow legs.

(d) *Blue*
Ground colour tones too light or too dark. A lack of lacing.

Breeding Hints

1. The minimum weights for breeders should be 4 kg for the male and 2.7 kg for the female.

2. Select females of good type and size.

3. Whilst a good fronted, low set cockerel is an attractive fowl in the show pen, he has physical problems in the breeding pen. The better bird in the breeding pen is a deep bodied bird showing plenty of shank. It may be necessary to crutch the feathers around the vent, with the preferred method being to pluck them rather than cut them with a pair of scissors as the feather stubs may cause pain or injury during mating.

4. When breeding Buffs, look for the following features:

(a) An even golden buff
(b) Make sure the breast tones with the other parts of the fowl
(c) The male must have sound buff wing and tail feathers.

Carefully observe the colour of Buffs throughout the year and select the ones that appear more colour fast. Avoid the ones that change colour after moulting.

Pekin

Breed History

The Pekin is a genuine or natural bantam with no large fowl equivalent. Its ancestors were imported into Great Britain by soldiers returning from the 1860 Anglo-French sacking of the Imperial Palace of China. The original stock were mainly of the Buff colour but some Blacks also found their way into Great Britain. Because the imported stock soon became inbred, crossings were made with other bantams of the time such as the Nankin to improve the general constitution of the birds.

Today the British Poultry Standards recognises the following colours: Black, Blue, Buff, Cuckoo, Mottled, Barred, Columbian, Lavender, Partridge and White. There are many other colours raised throughout the Pekin world and in particular those based on the Game fowl colours, for example Brown Red and Birchen.

Despite being distinct from the large Cochins, some overseas countries called them 'Cochin' bantams. This applies in Canada, USA, France, Holland and Germany.

The original attraction of the breed still holds true today, that is, Pekins do not wander or fly, can be confined by a 60 cm or 90 cm fence and respond well to handling. Because of the extensive footings, they must be kept on good clean, dry litter.

Because of their quiet, friendly nature and response to handling, Pekins have proved to be popular over the years. It is probably because of this popularity that the general standard of the breed has improved whereas others have fallen by the wayside. A good Pekin is always a top contender for major softfeather awards at any poultry show.

Positive Features to Look For *(colour plates p.92)*

The overall impression is one of a 'globe' or 'sphere', so that the fowl shows roundness, breadth, a 'ball-like' appearance, depth and width. From a side view, it should fit into a circle and just about touch all sides. Perhaps, the second impression to strike the observer is the profusion of feather. Interestingly, full adult plumage is not reached until the second year in many males and this quite often applies to many females as well.

The four main features of type that have to be looked for in Pekins are as follows:

1. *Correct shape*
As indicated already, Pekins should be globular and as close to the ground as possible.

2. *Correct carriage*
The carriage of the body should be as low as possible but not be a creeping gait. The carriage should be level

with a general inclination of the body to tilt forward. The top of the head should be just a shade above the top of the tail.

3. *Quality and quantity of feather*
There should be plenty of soft, pliable feather showing plenty of width.

4. *Abundance of fluff*
This is an extremely important feature of Pekins, as the larger the amount of underfluff, the more the feather will rise and sit out from the body giving the bird its characteristic soft feel.

Once due attention has been paid to these type features, the fancier can then look at the colour of the variety but without the correct type, we do not have a true Pekin.

The following is what we should be looking for in the individual parts of the fowl:

Head: Relatively small
Beak: Short, stout and slightly curved
Comb: Relatively small, single and finely serrated
Neck: Thick and short
Breast: Well rounded and carried well forward
Back: Short, broad and increasing in width to the cushion. The back should start to rise immediately behind the shoulders
Saddle: Should show a steady rise forming a well filled saddle. Width across the saddle is very important
Wings: Short with the ends tucked up neatly
Tail: Globular shaped, just protruding from the end of the cushion. The true tail consists of twelve feathers which are soft quilled. In the male, there should be many fine, highly sheened side hangers
Legs: Well spaced apart, showing plenty of width between them
Foot and hock feathering: When viewed from underneath, it should form one unbroken semi-circle from the middle toe and the thigh. The middle toe needs to be well feathered to its tip. There needs to be profuse feathering on the thighs and below the hock joint. If the bird is properly hock feathered, there is not likely to be a deficiency in toe feathering. The foot feathers should be carried at right angles to the legs.

Negative Features to Avoid

Here is a list of faults to avoid in the order that they would most probably be observed on a bird and not in order of importance. All should be avoided in Pekins.

1. Out of proportion birds, that is, not complying with the required spherical shape. This would include birds higher at the front than the rear, with long necks, longer than deeper bodied birds, birds with 'planted on' tails, birds with incorrect carriage ('creeper' or 'doormat' appearance) and those with a clean demarcation between foot-feather and body fluff.
2. Overly developed heads and eyes which are termed 'Mediterranean' heads in Pekin circles.
3. Nipped or pinched saddles instead of the curving line to the tail.
4. Tails where there is evidence of tail feathers being pulled out. Look for bent quills in main tail feathers.
5. Check the eye colour for pearl, hazel or dark eyed birds.
6. White in lobes.
7. Hard or stiff quilled feathers at the hocks. These are termed 'vulture hocks'.
8. Look at the wings and check for 'slipped' wings which are those of low carriage. In bad cases they will touch the ground. Open the wing out and check for 'split' wings in which there is a clear gap between the primary and secondary wing feathers. Pekins have been renowned for this fault which, once in a strain, is very difficult to breed out.

Breeding Hints

1. When selecting breeding stock it is not always possible to obtain the 'ideal' for the breeding pen, but the following should be looked for in the breeding stock. Firstly, birds as broad as possible between the legs. Secondly, extreme width of saddle. Thirdly, plenty of thigh fluff. Fourthly, foot and hock feather. 'Breadth' is the key word, especially in the saddle, body and feather.

2. Birds for the breeding pen need to be trimmed as the profusion of feather prevents successful joining of the vents during mating. Trimming, in the female, consists of cutting the feathers of the saddle and tail plus some around the vent. The male would need the 'footings' removed and some vent feathering. This would render both the male and female unsuitable for showing. An alternative to this is to use artificial insemination. This technique is used by most top breeders who can then still show their top birds. Artificial insemination techniques have been described earlier in this book.

3. Some people become worried by the colour of the chickens when they hatch as they bear no relation to

Mottled Pekin bantam female. Best Mottled, Pekin Club of N.S.W. Annual Show 1988. Owner: B. Treloar.

Black Pekin bantam male. Pekin Club of N.S.W. Annual Show 1988. Owner: W. Wingett.

Black Pekin male. Bantam Club of N.S.W. Annual Show 1988. Owner: F. Catt.

Cuckoo Pekin bantam male. Best Cuckoo Pekin, Pekin Club of N.S.W. Annual Show 1988. Owners: D. & J. Leyshon.

Splashed Pekin bantam female. Best Splashed Pekin, Pekin Club of N.S.W. Annual Show 1988. Owners: K. & G. Lambert.

White Pekin Bantam female. Pekin Club of N.S.W. Annual Show 1988. Owners: D. & J. Leyshon.

Black Pekin bantam female. Grand Champion of Show, Pekin Club of N.S.W. Annual Show 1988. Owner: P. Smith.

Black Pekin female. 1988 Sydney Royal Easter Show. Owner: Peter Smith.

Birchen Pekin bantam male. Pekin Club of N.S.W. Annual Show 1988. Owner: F. Fogarty.

Birchen Pekin bantam female. Best Birchen, Pekin Club of N.S.W. Annual Show 1988. Owner: F. Fogarty.

Cuckoo Pekin male. Fairfield Poultry Club Autumn Show 1988. Owner: Brian Dukes.

Cuckoo Pekin female. Fairfield Poultry Club Autumn Show 1988. Owner: Brian Dukes.

the adult colour. Culling on chicken down is to be avoided; wait until the true feather colour shows through. To give some idea of the problem, here are some of the down colours of some of the varieties of Pekins:

Whites: Pale yellow to pigeon blue
Buffs: Grey to deep cinnamon
Blacks: Black and yellow or white
Partridge: Brown with dark stripes.

4. Culling should be carried out retaining stock showing the following points:
Best carriage
Greatest width and broadest saddles
Broadest feather
Profusion of fluff.
Cull out those birds with sharp rising tails or cushions.

5. Assuming that some likely youngsters have been bred, a deal of work needs to be done to present the

fowl at its best in the show pen. The footings need a deal of 'work' done on them. This involves working the feathers so that they do not bunch, but instead slightly overlap each other as do the wing feathers.

The second concern is feather condition. Birds should be fed well and kept scrupulously free of external parasites. A good wash in a quality shampoo several days before showing will enhance good feather quality. The birds should then be kept sheltered on clean, dry litter.

6. *Colour breeding*
(a) Whites
Whites breed true so exhibition males and females can be bred from the one pen. Particular attention must be paid to type. Avoid any brassiness or creamy colour in the surface colour. Remember, they must be shaded or they tend to go yellow. Use clean white wood shavings as deep coloured wood could stain the footings if the shavings became damp.

If the colour is poor, use a Black male to the best White female and the best resulting cockerel back to the White hen.

(b) Black
A separate cockerel and pullet breeding pen is required to breed top exhibition stock. The cockerel breeding pen would be made up of an exhibition male mated to a female from a cockerel breeding strain showing dusky leg colour; but look for evidence of yellow showing through. In the pullet breeding pen an exhibition type female showing a dull matt colour is mated to a male from a pullet breeding strain showing the following features: yellow legs, brilliant green top colour, light colour in the fluff but not the quill and possibly a touch of red in the neck hackle.

There are some problems when breeding highly sheened birds together over a number of seasons as undesirable purple and bronze tinges start appearing in the plumage of the offspring. The bronze tinge can be counteracted by using a dull plum shaded female, the desired green sheen returning to the offspring.

A greenish legged female can also be useful for maintaining the leg colour in the males, but once again look for evidence of yellow under this green colour.

As can be seen, it is often necessary to keep a number of birds, not of exhibition type, but necessary to maintain the desired exhibition features. Room has to be made for the keeping of these fowls if a breeder is to continue producing top exhibition birds year after year.

(c) Buff
The Buff Pekin attracts the beginner with its soft golden shade but the variety is very difficult to breed to exhibition colour standard. On top of this, the type is often lacking as breeders have been more interested in colour breeding at the expense of type.

The colour ranges from a lemon shade to a reddish shade but it is the soft golden shade that is desired in the show pen.

It is claimed that double mating is not required with this variety. Careful attention though is needed to the breeding stock.

Male
An even colour is required and careful attention to the following:

(i)   No white, black or peppering in the tail or flights.
(ii) The undercolour should be sound, that is, buff to the skin and a buff quill. Especially look at the base of the hackle and tail feathers.
(iii) If possible, use a male that is known to be colour fast from moult to moult.
(iv) Watch for white in secondaries. This is a bad fault and the bird is probably a son of a mealy winged female. Mealy means white specks on the buff colouring.
(v)   Avoid birds with red hackles and saddles.

Female
Once again, good type is required coupled with an even colour. Slight colour compensations can be made but avoid using colour extremes or 'mottled' and 'mealy' offspring will result. Watch out for 'mealiness' on wing bows and lacing on breast feathers.

(d) Blue
These follow the typical Andalusian Blue pattern of incomplete dominance of colour, that is, Blue to Blue produce 50% Blue, 25% Black and 25% Splashed White. When a Blue bred Black and a Blue bred Splash are mated, 100% Blues result. Black to Black produces 100% Black, as Splash to Splash produces 100% Splash.

Attention should be paid to the blue colour, as it is a plain blue that is required not a laced blue.

(e) Partridge
Double mating is required here. In the cockerel breeding pen, an exhibition male is mated to a female with pale neck feathering and light pencilling. The pullet breeding pen consists of an exhibition female with good ground colour and markings mated to a male with dark top colour, coarse hackle markings and broken breast.

Irrespective of the colour chosen, the first priority must be type and secondly colour. Combine the two and you have a top exhibition fowl, the envy of all fanciers.

Polish

Breed History

The Polish (Poland) breed is a very old breed of fowl being mentioned as a pure breed as far back as the sixteenth century. It must have been imported into Great Britain a long time ago as it was exhibited at the first poultry show in London in 1845 and was described in the first Standard produced in 1865.

Great Britain has standardised several colours: Chamois (White Laced Buff), Gold, Silver, White, White Crested Black, White Crested Blue, White Crested Cuckoo. Bearded varieties are also recognised. Overseas the Polish is a highly prized ornamental breed, but there are few in Australia. The breed exists in both large and bantam form.

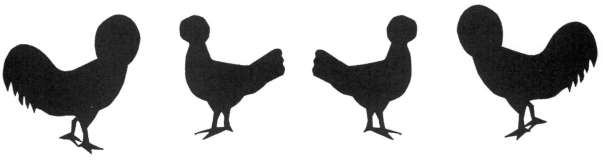

White Crested Black Polish. Camden Poultry Club Annual 1988. Owner: B. Raines.

White Crested Black Polish. Camden Poultry Club Annual 1988. Owner: B. Raines.

White Crested Black Polish large fowl female. 1988 Sydney Royal Easter Show. Owner: Bruce Raines.

Positive Features to Look For

1. Poll or crest: A globe of feathers placed firmly on the skull in a balanced fashion
2. Carriage: Alert
3. Body: Compact and rounded on sides
4. Back: Medium length with the sides tapering to the base of the tail
5. Breast: Bold, rounded to balance the crest on the head
6. Shoulders: Wide
7. Wings: Carried well up
8. Neck: Gives the impression of rising from the shoulders or a little further back than other breeds
   Wide appearance to balance the crest on the head

Negative Features to Avoid

1. Most attention tends to be placed on the crest. Common faults include those that are
   (a) split in appearance
   (b) twisted
   (c) lopsided
   (d) out of shape, that is, oval instead of round when viewed from the front, side or top.

2. In the laced varieties, crescent shaped markings present instead of the desired lacing.

3. What could be regarded as physical or structural deformities.

Breeding Hints

The general comment on Polish is that too much emphasis is placed on the crest at the expense of size and in some varieties, the lacing. The crest or poll must be ball shaped. Those well versed in the breed claim that the male has the greatest influence over this feature, so special care must be taken in selecting the male to head the pen. This does not mean that females should not be carefully scrutinised. The best results are achieved when this feature is strong on both sides of the mating. Even in the best matched pens there will still be culls, but this is a better approach than relying on 'fluke' matings.

Plymouth Rock

Breed History

When most people hear the breed Plymouth Rock mentioned they picture the attractive Barred variety laying brown eggs and providing a sizeable table bird. It has established itself, along with the Rhode Island Red, as arguably the finest of American dual purpose poultry. Its attractiveness has sustained several generations of breed supporters in many poultry fancier families.

Plymouth Rocks are commonly believed to have a background of Dominique and Black Cochin or Black Java blood. Barred Rocks from this ancestry were first exhibited in 1869 and so were brought to the public's notice as a new breed in the poultry world. Their commercial value was quickly recognised and popularity rose.

The breed reached Great Britain in 1871 and soon established itself there. Barred Rocks further came to prominence with the work on autosexing breeds at Cambridge University, following the development of the Cambar breed, when Punnett and Pease crossed the Barred Plymouth Rock with the Campine.

When the breed arrived in Australia is not clear but it was certainly present and eagerly supported in the 1890s. Of recent times many have placed undue emphasis on the barring at the expense of size and stamina. This has led to a general decline in the quality of the breed. However, the recent establishment of the Plymouth Rock Club of Australia and the enthusiasm of its executive should see this situation reversed. The Club has taken the step of publishing its own Standard by which it would like the breed to be assessed in this country. The Standard recognises the following varieties, Dark Barred, Light Barred, White, Buff, Columbian, Silver Pencilled, Partridge, Blue and Black. Bantams are also recognised in the above colours.

Positive Features to Look For

The type to look for can be summarised as follows:

1. Body: Long and straight, broad and deep
2. Breast: Well rounded and prominent
3. Back: Long, straight, horizontal
4. Wings: Well tucked up and not excessively long
5. Tail: Carried at about 30° to the horizontal, well formed and of medium size
6. Neck: Reasonably long, well arched, plenty of hackle
7. Head: Medium sized
8. Eyes: Large and prominent
9. Beak: Relatively short, curved
10. Comb: Single, reasonably small, five serrations preferred
11. Wattles: Medium sized, fine texture.
12. Ear lobes: Smooth oblong
13. Legs and toes: Medium length, plenty of space between thighs.

The female, apart from the usual sexual differences, should be similar to the male.

Negative Features to Avoid

Features to be avoided or culled out include the following:

1. In knees or weak hocks
2. Slipped wing
3. Crooked or twisted toes
4. Excessively leggy birds
5. Cut away fronts
6. Twisted wing feathers
7. Slow feathering strains
8. Sunken eyes and faces
9. Split tails in males
10. Two or more solid black feathers in the primary or secondary wing feathers or the main tail
11. Red or yellow in barred plumage
12. Poorly defined barring in the barred variety.

Breeding Hints

1. Barred

As has been mentioned previously under breed history, the Barred variety in general has suffered from a disproportionate emphasis on the barring at the expense of breed type. While it can be argued that a Barred Rock is a 'nothing' fowl without its barring, the breed totally loses its identity if not enough attention is paid to the actual breed type that distinguishes it from all other breeds. This has been reflected in Barred Rocks with a reduction in size and loss of front or breast. While what follows concerns itself with barring, the breed type must not be forgotten.

Three points must be remembered with the barring. Firstly, the barring should go as far down the feather as possible. Secondly, it should be straight across the feather. Thirdly, the barring must have a sharp edge to it and not be fuzzy or indistinct.

The Plymouth Rock Club of Australia now recognises two forms of barring, light and dark. The fowls in question used to make up the old pullet and cockerel breeding pens prior to the release of the Club's Standard.

(a) Light Barred

Formerly known as the pullet breeding pen, this pen is made up of a male from a known pullet breeding line with a white bar double the width of the dark bar on his plumage, and a female showing barring that is of equal width. Look carefully at the barring on the female's breast particularly the lower breast where it

Barred Plymouth Rock bantam female. 1988 Sydney Royal Easter Show. Owner: J.T.F. Wong.

Large Dark Barred Plymouth Rock male. Fairfield Poultry Club Autumn Show 1988. Owner: J. Gardner.

Large Dark Barred Plymouth Rock female. Fairfield Poultry Club Autumn Show 1988. Owner: J. Gardner.

tends to be weak. Both male and female of this pen can now be exhibited in the Light Barred Rock classes where they are provided at shows.

(b) Dark Barred

This pen was formerly known as the cockerel breeding pen. It is made up of a male with barring as nearly equal as possible in width. Watch for lightness in the neck hackles. The female required needs to be cockerel bred, dark, with black barring twice the width of the white bar. Likewise, both male and female of this pen can be exhibited in the Dark Barred Rock classes.

What in essence has happened, is that rather than leave the cockerel breeding female and the pullet breeding male at home under the old Standard, these can now be legitimately exhibited in classes provided for them. The move does have the advantage that it ensures that cockerel and pullet breeding lines are kept separated, not mixed as often happens with other double mated varieties. Opponents of this move would say that it will needlessly increase the number of classes at shows, confuse the uninitiated and make classes for formerly nonstandard birds. No doubt much discussion will continue to be generated in poultry circles by this move.

2. White

There appears to be little information available on breeding Whites so it can only be assumed that they bred true.

3. Buff

Since the Buff variety was developed using Buff Cochin blood, it would be reasonable to expect this variety would share the colour breeding problems of the buff colour as seen in Leghorns, Pekins and Orpingtons. Reference to the breeding hints for those varieties would assist with breeding Buff Rocks.

4. Silver Pencilled

Developed by using Dark Brahma and Silver Pencilled Wyandotte blood. Refer to breeding Silver Pencilled Wyandottes to assist with this variety.

5. Partridge

Once again, check with the breeding of Partridge Wyandottes for advice here, as these were used in producing the variety.

6. Columbian

The colour pattern closely follows that of the Columbian Wyandotte so refer to the breeding of the Columbian variety of that breed to assist with breeding Columbian Rocks.

While specialist breeders have developed other colours the Barred variety remains the premier variety of this breed. The natural attractiveness of this variety will probably ensure that it keeps its position.

Rhode Island

Breed History

Many claim Rhode Island Reds were the greatest utility fowl in the world, both good egg layers and excellent table fowls. This claim may well have been true until the advent of the modern commercial hybrid fowls which exceeded the Rhode Island's egg laying ability and growth rate as table birds. The breed still finds much favour amongst hobby farmers but their selection of the breed is probably based more on sentimental judgement than objective measurement.

The breed took its name from the State in the USA where it was developed as a breed in the late 1800s. The common red fowl of the area, the Chittagong Red, was crossed with Cochin, Malay, Brahma and Brown Leghorn blood to develop the Rhode Island as we know it today. The first Standard was drawn up in 1901 and the breed admitted to the American Standards in single comb form in 1904 and rose comb form in 1906. The breed was soon exported to Great Britain where it became an instant success.

Rhode Islands are seen mainly in the Red colour, however, following crossing with Partridge Cochin, White Wyandotte and rose combed White Leghorn, a White variety was produced in the State of Rhode Island and admitted to the American Standard in 1922. It is a little surprising that the White variety did not become more popular, as it had all the utility properties of the Red variety without the hassles associated with breeding the Red colour.

Rhode Island Red bantams were produced soon after the large fowl was standardised and still remain reasonably popular amongst certain sections of the bantam fraternity.

The first Rhode Island Red large fowls were imported into Australia by J.R. Dalrymple of Bexley, N.S.W. in 1912. The breed caught on rapidly with large classes seen at the Sydney Royal Easter Show in the years that followed. The Rhode Island Red has been used commercially in Australia but not to the same extent as the White Leghorn and Australorp. The breed proved popular up to the 1930s, then declined, with a revival of interest in the late 1950s before it was overtaken by the modern commercial fowl. One problem that has plagued the breed is that it has suffered at the hands of 'colour faddists' who were more interested in attaining the Standard colour than maintaining the breed's inherent utility properties.

The first Rhode Island Red bantams were imported from Great Britain around 1927–28 by E.W. Jones of Hurstville, N.S.W.; however, he made little progress with them and they were passed on to the legendary C.A. Clark at Baulkham Hills. Many hundreds of kilometres away at Tully in Queensland, Arthur Harwood was busy breeding down a strain of Rhode Island Red bantams from selected large fowl. This strain proved superior to the imported strain.

Today a good Rhode Island Red is rare, but when they are produced they are capable of winning major show awards.

Positive Features to Look For *(colour plates p.93)*

As has already been mentioned, in the past far too much attention has been paid to colour to the detriment of type. Despite the fact that 20–25 points are allocated to colour and plumage, the first consideration must be type. If not, the very essence of the breed is lost.

The usual way of describing a Rhode Island's body shape (type) is as 'brick shaped'. This can be put two ways. Firstly, hold a brick side on along the body of the fowl. The body of the fowl should follow the outline of the brick. Secondly, imagine two bricks, side on, stacked on top of each other. The top brick would represent the fowl's body and desired carriage. The space occupied by the second brick would represent the space from the ground to the underside of the keel.

Turning our attention to specific parts, we should be looking for the following:

1. Body: Oblong, brick shaped
2. Comb: Medium sized, of fine texture with five even serrations
3. Wattles: Medium length of fine texture
4. Lobes: Red
5. Beak: Short, strong and yellow or horn in colour
6. Eye: Red and filling the eye socket
7. Neck: Relatively small, arched and, in the female, with black ticking on the hackle feathers at the base of the neck
8. Breast: Rounded, broad and deep
9. Back: Broad, long and nearly horizontal
10. Wings: Carried well up and close to the body
    Black markings on the lower web of the primaries and black markings on the upper web of the secondaries
11. Tail: Rising only slightly from the flat back line. Brilliant black–green sheened tail feathers
12. Legs: Yellow with a line of red down the side
13. Skin: Yellow
14. Plumage: Feathers red to the roots
    Male's neck hackle clear.

Negative Features to Avoid

Firstly, we shall turn our attention to the physical faults associated with type:

1. Excessive size: This is a utility breed and birds of excessive size are generally not productive
2. Carriage: Should be level and not inclined
   Avoid excessively leggy bantams
3. Comb: Side sprigs
4. Wattles: Long folded type

5. Ear lobes: Coarse, thick type and any showing permanent white in them
6. Eye: 'Walled' eyes that do not fill the eye socket
   Greenish grey coloured eyes
7. Backs: Crooked backs or those showing raised 'kidney' bones when the hand is run down the back
8. Wings: Drooped or slipped wings
9. Tails: Wry tails and long streamy sickles
10. Legs: Flesh coloured legs instead of the required yellow.

Secondly, turning our attention to the vexed question of colour in Rhode Island Reds.

1. Top colour: Brown, chocolate or multicolours
   Black on the shoulders
   One or more entirely white feathers showing in the outer plumage
2. Undercolour: Very pale or smutty, sooty under-fluff, mealiness
3. Neck hackle: Gingery or yellowish necks
   Too light on surface of neck ('pumpkin neck')
   Heavy lacing in the hackle of the female
   Black in the male's hackle
4. Breast: Yellowish shaded breasts
   Lacing or shaftiness on the breast
5. Wings: White or grey showing in the wings
   Flights with grey tips or excess of black
   Red in the primary and secondary feathers
   Not matching the surface colour
5. Tail: Leghorn type tails
   Bronze colour in tails
   Grey colour at the base of tails.

Breeding Hints

The keys to breeding this breed are firstly, level carriage in type and secondly, evenness in colour. It cannot be too strongly emphasised with this breed that type is the number one priority and must be seen to first, then attention to colour.

In breeding Reds to exhibition colour, the theme is 'even colour all over'. When selecting breeding pens, try to put together fowls of good colour and type first. Other matings are regarded as 'chance' matings where there may be a possibility of gaining some good coloured progeny but there will also be lots of culls. Normally, well coloured birds will breed true for several seasons, then things degenerate. The following is what can be done in such situations:

1. If the male carries too many black markings or smut in the undercolour, mate him with a female clear winged and lightly marked.

2. If the male is lacking in undercolour or markings, mate him to a female somewhat heavy in black points. Remember the birds in both these matings must be of good type.

'Smut' or 'soot' can be used to counteract the lack of richness in the top colour, but an excessive amount of 'smut' or overuse of 'smut' will produce excess ticking in the offspring. The desired amount of 'soot' or 'smut' is a sooty line in the undercolour showing as a bar across the feather on the back just past the shoulders.

Too much emphasis on black points in the female could cause the resulting cockerels to lose their desired clear neck hackle. It is better to pay more attention to the maintenance of the black points in the tail.

White variety
The white variety breeds true for colour with the occasional need to cull based on 'non-white' feathers appearing in the plumage. As with most white fowls, show specimens should be shaded to achieve the snow white plumage needed.

Rosecomb

Breed History

The Rosecomb is a true bantam with no large fowl equivalent. They are purely an exhibition fowl and, at their best, are a real show stopper. Their ancestry is believed to extend back to the reign of Richard III who is said to have taken an interest in native Rosecomb-like fowls in 1483. Rosecombs are well documented in poultry literature from 1787 onwards, so it can be assumed that this breed has given much pleasure to poultry fanciers over many centuries.

Today the British Poultry Standards recognises three varieties, Black, Blue and White. Likewise, the American Poultry Standards recognises the same three varieties, Black and White being included in the first Standard published in 1874. Interestingly, the American Bantam Standards recognise 26 varieties of Rosecombs, most based on Belgian and Game colours. Others include Buff, Black Tailed Red, Buff Columbian and Exchequer.

According to information available to the author the Rosecomb breed was present in Australia pre 1890. They have proven very popular amongst Australian fanciers, even back in those times when lots were imported from Great Britain with high prices being paid for selected imports. Today there is an active breed club which gives its members much support.

Positive Features to Look For

The type to look for can be summarised as follows:

1. Carriage: Proud and stylish, often described as jaunty.
2. Body: Short, broad, wide at the shoulders, cobby build.
3. Back: Short and flat with a blunt end to house the flowing tail in the male.
4. Breast: Rounded, carried well up and forward.
5. Wings: Wing carriage is important as they have to be carried just off the body and low enough to only reveal the front of the thighs.
   The wing feathers should be well rounded.
6. Saddle: In the male, well filled out with feathers.
7. Tail: Gently rising angle to the back, made up of broad, overlapping main tail feathers.
   In the male, the sickles long and broad with rounded ends.
   The side hangers should also curve to complete a half circle and fill in the side of the tail.
8. Thighs: Set well apart.
9. Shanks: Stout.
10. Toes: Straight and well spread.
11. Comb: Short, broad rose with plenty of fine parts ('workings').
    Set squarely on the skull and level.
    Leader tapering to the rear end, straight and slightly away from the neck line. The female comb much smaller but of similar form.
12. Lobes: Round, circular edges and lying flat on the face.
    Purest white with a kid-like texture in the male; the size should be 19–22 mm or approximately in the range of a 5 cent and 10 cent coin.
    In the female, the size required is 16 cm or about the size of a 1 cent coin.
13. Face: Fine textured and cherry red in colour.
14. Eyes: Quick and bold in appearance, hazel to brown in colour.
15. Beak: Small.
16. Feather quality: Broad in width, must show a healthy sheen.

Negative Features to Avoid

With a breed that has reached a high level of perfection, the list of negative features to avoid or cull out is quite lengthy. None the less if a breeder wishes to reach the top there is no room for inferior birds as the competition in this breed is intense. Features to be avoided or culled out include the following:

1. Dumpy bodied birds.
2. In the males, tail feathers protruding past the sickle feathers.
3. Bluish lobes.
4. Poor wing carriage, either wings carried too far up or too drooped.
5. Narrow fronted birds.
6. Gypsy (mulberry coloured) faces.
7. Any comb defects, such as oversized combs or poorly set combs.
8. Look for lobe faults such as mis-shapen lobes, dished lobes, hollow lobes, folded or creased lobes. Cull out oversized lobes except where they are specifically being used in the breeding pen.
9. Leader problems such as those that are too thin or droop.
10. Stiff legged birds or those too far up on their legs.
11. Coarsely boned birds.
12. Light red or yellow eyes.
13. Pale leg coloured birds.
14. Coarse faced birds such as those whose lobes do not sit properly on the face.
15. Feather faults to be on the look out for include:
    (a) Steely coloured patches in flights
    (b) White tips
    (c) Bronze sheen in the Black variety
    (d) Purple sheen
    (e) Lack of width in feather—very important.
16. White in the red face especially in pullet breeding males.

Breeding Hints

1. Perhaps the biggest danger in breeding Rosecombs is to fall into the comb–lobe–tail syndrome at the expense of the rest of the bird. To be lured into this is somewhat understandable as these three features make up 50 out of the 100 points allotted, however the tendency should be resisted.

2. Cockerels tend to remain fresh and sappy for a few weeks then start to decline and lose their bloom. Top breeders hatch chickens at strategic times to ensure they have cockerels at their peak for important shows.

3. Blacks
While some breeders are successful using a single pen to breed winners of both sexes, most top breeders rely on a double mating system to produce top exhibition birds. As with some other breeds, for example the Leghorn, it is very difficult to produce males with the

Black Rosecomb bantam female. 1988 Sydney Royal Easter Show. Owner: Alex Lamont.

Black Rosecomb bantam male. Camden Poultry Club 1988. Owner: A. Cheetham.

desired flowing tails and exhibition females from the one pen.

(a) *Cockerel breeding pen*

The following female is the one needed to be matched to an exhibition male to produce exhibition cockerels. She must have:

(i)   A good well 'worked' comb
(ii) Good length of leader
(iii) Thick round lobes that sit squarely on the face
(iv) Wide main tail feathers with the top two slightly arched
(v)   Plenty of cushion feather and tail coverts
(vi) Short back
(vii) The sheen should be a little dull, perhaps described as matt black or coal black
(viii) Body carriage must be confident.

(b) *Pullet breeding pen*

To produce top pullets, use second year hens with sound lobe, good colour and wide tails. The pullet breeding male needs to have the following features:

(i)   A comb larger than the exhibition male but full of fine 'workings', as pullets often lack comb size
(ii) Lobe which is slightly larger than the standard
(iii) Long leader
(iv) Correct type with a short wide body
(v)   Wide feather
(vi) Less tail development than exhibition male, well fanned and carried high
(vii) Grey in the undercolour at the base of the tail
(viii) A little red in the neck and saddle hackle is useful in producing a good sheen in pullets especially if the female used is of the matt black shade of feather.

4. Whites

The White variety appears to breed true to colour but seems to suffer from poor lobe and feather with prolonged white to white matings. These features can be freshened up using a Black Rosecomb in the following way:

(a) Cross with a desirable Black Rosecomb.
(b) The progeny of this cross will be all Black.
(c) If selected members are crossed, Whites will be produced in a ratio of 1 White to 3 Blacks.
(d) From this point on, the Whites will breed true.
(e) If the White offspring of the White–Black cross show blue in the legs, these can be mated to pure Whites and the leg colour will be restored.

5. Before leaving breeding and producing desirable Rosecombs, some mention should be made of looking after the lobes and combs.

(a) Keep exhibition stock in well shaded and protected pens.

(b) An old time recipe for treating blistering of lobes is as follows. Treat the lobe with a mixture of lemon essence and water at the rate of two drops of essence in a tablespoon of water. Carefully dry the lobe and touch up with glycerine or dust with starch powder. Keep the fowl in a semi-darkened or well shaded pen.

(c) Combs respond to a regular treatment of lanoline or zinc ointment.

If a stylish, fancy fowl appeals to you, then the Rosecomb is a hard breed to go past. The breed also has the back-up of a number of good breeders and an active breed club.

Sebright

Breed History

The Sebright is a genuine ornamental bantam that was carefully produced by Sir John Sebright after 30 years of painstaking work around 1800. The breed is different to others in that the male is henfeathered, that is, having no neck hackles, saddle hackles or sickle feathers. To compensate for this, each of the feathers is individually laced around its edge with green–black. The ground colour is either white or golden bay giving the two varieties of Silver and Gold respectively. The breed also has the distinction of having the first recorded specialty club formed in 1815 to foster the breed.

Sebrights were also used in the development of the Laced varieties of Wyandottes in America. The breed was recognised in the first American Poultry Standard of 1874.

The author has no record of when they arrived in Australia but evidence shows that they were here pre 1900.

Positive Features to Look For *(colour plates p.94)*

Much has been said about the Sebright type which is quite distinct from other breeds. Some have likened it to a fantail pigeon but this is probably a little extreme. The main type features are as follows:

1. Carriage: Strutting, cheeky.
2. Stance: Gives the impression of standing on its toes.
3. Body: Broad and compact.
4. Breast: Prominent and full.
5. Back: Short, flat and ending squarish rather than the round taper to the tail.
6. Wings: Long and carried low.
7. Tail: Carried high at about 70° to the back, large and well spread.
   No sickles or side hangers in the male.
8. Legs: Set well apart. Slate blue in colour.
9. Neck: In the male the neck is arched and carried well back.
   In the female carried upright.
10. Comb: Rose, sitting squarely on the head.
    Leader extending slightly upwards at the back, more like a Wyandotte than a Rosecomb.
11. Comb, face, wattles, ear lobes: Dark purple or mulberry preferred but mostly seen as a dull red. Hens tend to have darker faces.
12. Plumage: Short, tight and almond shaped lacing extends from the throat to the thighs.
13. Lacing: Beetle green black in colour.
    Flights, tail, throat and shoulders must be evenly marked.

Negative Features to Avoid

As with many ornamental breeds, quite often far too much emphasis is placed on a single feature instead of the whole bird. Sebrights have suffered from 'lacing' obsession at the expense of type and other distinctive breed features, such as the mulberry coloured faces that have all but disappeared from Sebrights today. Whilst the comments that follow do place heavy emphasis on lacing, the rest of the bird should not be forgotten. Look carefully for the following:

1. Variations in ground colour, either too dark or too light need to be rejected. Look also for creamy ground colour in Silvers, and buff tones in the Gold variety. Grizzliness in ground colour needs to be culled out as well.

2. The following problems can be found in the lacing:
(a) Lacing that goes brownish black or grey on the edges.
(b) Frosty or shadowy lacing.
(c) Uneven, broken, non-continuous lacing particularly on the big feathers. Sometimes the lacing runs larger on the ends of the big feathers.

3. Single comb sports.

4. White in the earlobes.

5. Spotty and dusky markings around the base of the tail feathers. Peppery feathering quite common in tails.

6. Ultra short backs and short legged birds.

7. Transparent ground colour, particularly in the Silvers where the underlying lacing shows through the Silver ground cover.

Breeding Hints

To breed Sebrights to Standard presents a challenge with rich rewards of satisfaction and admiration if a good one is produced. Here are some points which may help achieve that goal.

1. Sebrights are not good layers and are generally late starters in the spring.

2. They are not a robust breed and fertility is often poor. Some feel that this is due to the male being hen feathered and suggest that a male with slight sickle development, say about 2 cm past the tail feathers, be used. Another suggestion is that this breed could respond to the use of artificial insemination to overcome the henfeather problem.

3. There is no need to double mate as Sebrights breed true provided due attention is paid to type and lacing.

4. The desired lacing is achieved by either matching two birds of medium width lacing or by balance mating, using a widely laced bird with a thinly laced bird. Pay particular attention to the lacing on the wings, wing butts and tail as weaknesses frequently appear there.

5. Do not cross the Gold and Silver varieties or the ground colour of both will be ruined in the offspring.

6. Protect growing stock from the weather as the Silver ground colour will go brassy and the gold bleaches out.

7. If double lacing appears, carefully check both parents for this fault, then eliminate the offending bird(s) from the breeding pen.

8. Because Sebrights are henfeathered, care must be taken in putting the pens together if more than one female is used with a male. Assertive females have been known to almost scalp a less dominant male.

Despite the problems, the Sebright presents a breeding challenge to stimulate even the most experienced poultry fancier. There is no reason why they could not also be taken up by a keen and alert beginner.

Silkie

Breed History

Silkies are an Asian breed but there is some debate as to where they did actually originate. Japan, China and India have all been suggested as places of origin. The breed is an ancient one and was mentioned by Marco Polo in literature dating back to 1298.

Silkies were admitted to the British Standard in 1870 after arriving in Great Britain around 1860. Today they are recognised in Black, Blue, Gold and White but not in their bearded form. In Great Britain they are regarded as a light breed of large fowl.

Across the Atlantic in the USA they are classified differently. They are regarded as bantams with only two colours being recognised by the American Poultry Standards which also recognises the bearded variety in both colours. The American Bantam Standard, however, recognises 10 varieties, the bearded and non-bearded versions of Black, Blue, Buff, Partridge and White.

The breed's distinctive plumage is the obvious attraction. The feathers lack the normal barbs which hold the webs of the feather together producing long masses of 'silver threads' instead of the usual feather vane. As a result the breed does not fly and can be easily restrained in an area. People find Silkies very easy to handle, and they become extremely friendly towards their keepers. Coupled with this, they are also very hardy and adaptable.

When crossed with the Wyandotte breed excellent broodies are produced, widely spoken of as the best by those who use broodies to hatch eggs. Further, if the cross is made using a Silkie male and a White Wyandotte female, the offspring show sex linkage in their leg colour and comb colour. The males have dark legs and dark faces, whereas the females have white legs and faces. The pure Silkie is renowned for its broodiness too, but the cross is preferred as there is less chance that chickens will strangle themselves on the long silky plumage.

Positive Features to Look For *(colour plates p.94)*

Despite the unusual apearance of this breed, there is a definite type which makes up a Silkie. The things to look for are as follows:

1. Carriage: Bright, alert and attentive.
2. Body: Ball-like body, broad and square at the shoulders.
3. Back: Short.
4. Breast: Broad and full.
5. Tail: In the male the tail feathers have a ragged or frayed appearance and should rise from the saddle in a short round curve.
   In the female the tail is almost hidden by a dense saddle (cushion).
6. Thighs: Well covered with fluff.

7. Legs: Reasonably short and slightly feathered.
8. Feet: Five toes should be clearly visible.
   The feet almost covered with fluff.
   Middle and outer toes feathered.
9. Crest: Dense, silky and streams gracefully over the neck of the male. Made up of about 12 feathers, the side ones not obscuring the vision of the bird. In the female the crest is ball shaped squarely balanced on the head.
10. Comb: Pea comb, large in the male and frontally broad with workings only slightly above the comb surface. Much smaller and less prominent in the female.
11. Head: Short.
12. Beak: Slaty blue.
13. Eyes: Black and bold.
14. Comb, face, wattles, skin: Mulberry.
15. Earlobes: Turquoise blue.
16. Plumage: Silky, abundant, long and soft.

Negative Features to Avoid

The following need to be looked for and culled out or avoided if buying birds to breed with:

1. Hard feathered vulture hocks
2. Green beak, legs and pads
3. Red face or comb
4. Light eyes
5. Split or inadequate crest
6. Four toes
7. Excessively feathered shanks.

Breeding Hints

If you look at the points allocation for the Silkie breed, the priorities could be grouped as follows:

Head and plumage
Type
Legs and colour.

As has been pointed out before, it is important to place type first irrespective of points weightings.

1. When selecting birds for type, look for short legs, wide chests and a short back.

2. Specifically the male needed in the breeding pen should have good silk, abundant crest with no breaks, a short, compact, cobby body and be well furnished. The female should be short and cobby with a pompom-like crest with no splits.

3. Silkies should be protected from the weather, particularly the Whites which tend to go brassy.

4. Blue Silkies can be produced by crossing the Black and White varieties but care must be taken with the selection of the Black as this influences the type of Blue produced. Several Blacks may have to be tried before a satisfactory tone is achieved.

The Silkie has universal appeal amongst a large number of people of all ages. Close relationships can be built up with these fowls if their keeper is prepared to give them the time.

White Silkie female. 1988 Sydney Royal Easter Show. Owner: L.S. Huntingdon.

Bearded White Silkie female. Fairfield Poultry Club Autumn Show 1988. Owners: S. & D. Lindsay.

Sussex

Breed History

The Sussex breed is regarded as one of the older breeds of softfeather fowls with its origins having been traced back 175 years ago to the Sussex area of England. The fowls were of mixed appearance and favoured by the suppliers of poultry meat to the London markets. Their white flesh was greatly prized and sought after.

The Sussex Breed Club was founded in 1903 to promote the breeding, exhibition and standardisation of the breed. This club is still active in England today.

Speckled Sussex are regarded as the oldest member of the Sussex family followed by the Reds. Brahma, Cochin and Grey Dorking were used across existing Sussex fowls to make the Light variety and the infusion of such blood probably gives rise to some of the faults and disqualifications outlined in the current Standards.

The Light variety played an important role in the work on sex linkage, acting as a 'silver' in the gold and silver sex linked crosses. In fact a Light Sussex female crossed with a Rhode Island Red male has become the standard example of a sex linked cross.

Sussex large fowls found their way to Australia at the turn of the century with a documented importation of Lights, Reds and Speckleds in 1905. However, little progress was made until the 1920s when importations by J.E. Gibson set the breed on the road to popularity. Lines from this importation have been maintained by a few dedicated breeders to the current time.

Sussex were bantamised in England about 1934 and by 1936 were being exhibited in several colours mainly Lights and Speckleds. Whites, Reds and Browns followed soon after. Silver bantams appear to be a more recent creation.

Today the Sussex has found a niche as a farm fowl providing excellent table features and reasonable egg laying ability with good winter persistency.

Positive Features to Look For

1. Sussex fowls are a large sized fowl with no upper limit for weight specified in the breed Standard.
2. The back needs to be long, flat and broad.
3. They have a large round front indicating length of breast, an important meat quality. The large full front tends to be more readily seen in older females.
4. The body must show depth of frame through the back to the thigh region.
5. The body is carried on white shanks and feet of moderate length, set wide apart. Some favour a red pigment running up the sides of the shanks as they claim that this is an indicator of stamina. However, the breed Standard specifies white.
6. Handling is important as this reveals the bird's table qualities for which the breed is renowned.
7. The bird should be alert with a clean neat head, clean face and round orange or orange-red eye.

Light Sussex bantam female. Bantam Club of N.S.W. Annual Show 1988. Owner: G. Childs.

Light Sussex bantam male. Bantam Club of N.S.W. Annual Show 1988. Owner: G. Childs.

Large Light Sussex male. Camden Poultry Club Annual 1988. Owner: B. Raines.

Large Light Sussex female. Camden Poultry Club Annual 1988. Owner: B. Raines.

The eyes should almost be seen from behind the skull when viewed from that position. The comb should be of medium size and evenly cut.

8. The bantam type must be a miniature of the large fowl type with emphasis on the long, flat back and depth of body. The priorities should be firstly, type; secondly, pureness of colour; and thirdly, size. The male should be around 1100 g and the female 790 g, but these are often exceeded.

Negative Features to Avoid

1. Some pullets do not show their true size until they begin to lay but should be culled if under 2.5 kg at five months of age. To maintain size in the fowls, large females should be used in the breeding pen.
2. Cull out birds with poor fronts, for example, 'cut away' fronts.
3. Feathers or feather tufts on legs are a disqualifi-

cation and sometimes appear as a result of Brahma blood in the original make up of the breed, particularly the Light variety.

4. Reject leggy specimens.
5. Loose feathering can also be traced back to Brahma and Columbian Wyandotte blood being used in the breed's make up. Loose feathering gives a false impression of the bird's true type.
6. Cull out birds with heavy eyebrows.
7. Combs need to be free of side sprigs and not have a fold in the front over the top of the beak.
8. The lobes should be red and free from white. The fold in the comb and white in the lobe can invariably be traced back to the infusion of Leghorn blood to improve the egg laying capabilities of the breed.

Breeding Hints

1. Light

(a) It is important to remember that the size comes

from the female but the colour is largely controlled by the male. However, females with clean body colour, free of sootiness must be used in the breeding pen.

(b) Faults to be on the look out for when picking the breeding pens are:

(i)   Unsound tail colour in males particularly white in the tail.
(ii) Brown or brownish grey or grey hue in the black parts.
(iii) Yellow staining on the top of males.
(iv) Black feathers in the top colour or any part other than the neck or wings.
(v)   Grey or white in quills of hackle feathers, this being termed 'shaftiness'.

(c) The neck and tail markings appear to be closely related but the black in the wings appears to be less directly related to the necks and tails.

(d) 'Sooty' or 'smutty' males (light slate grey in the undercolour) can be used with clean bodied females as these males contain a certain reserve of colour. They should only be used in the breeding pen and not the show pen. Another indicator of colour reserve in males is the presence of ticking in the saddle hackles, but once again this should not appear in the exhibition male, only the breeding pen male.

(e) Careful attention must be paid to the neck which should have a clear black centre surrounded by a clear white border and should be free of dark edging or 'fringing', giving a lace appearance.

(f) A green sheen in the black parts should also be strived for.

(g) A clear black should be present in the flights but the wing should show as white when closed against the body.

(h) It is very difficult to get uniformity of colour markings across a flock of Light Sussex so there is a need for a process of levelling up or balanced matings to produce the desired individual for exhibition.

(i) Continued mating of exhibition stock leads to a loss of black colour giving the black parts a 'washed out' appearance or 'pencilly' neck hackles. This establishes the need for the 'sooty' male in the breeding pen to act as a colour reserve for when this situation arises.

(j) Early culling of cockerels can take place for such faults as too much black in the undercolour, badly striped saddle hackles and white in tails. These will not improve with time. Birds with deficient neck hackles can be safely culled at around four months of age.

2. Speckled

(a) In the breeding pen the males should be a bright rich mahogany in the ground colour, open in breast markings and broken on top. The dark mahogany ground colour produces bad white markings and heavy black necks in both sexes. The females on the other hand should be rich in ground colour with clean white markings.

(b) The feather colour is a tricolour pattern consisting of a rich mahogany colour with a narrow glossy black bar tipped with a white spot. The black must be free from smears of white or brown. The white tip must be snow white without specks or smears.

(c) Culling of young birds can begin early if the following show up. In the young cockerels, gingery colouration in the neck and saddle hackles, white in the undercolour which should be slate or red, breast feathers heavily splashed with white and too much white in the tail or wing bow. Any young pullets that are very washy or very black in the ground colour or surface colour, or have oversized white markings (too gay) should also be culled.

3. White
White is relatively straightforward but some breeders feel that a little black in the plumage enhances the pure white colour in the offspring.

4. Red
Whilst Reds are extremely rare in this country, the following may be of help to those who care to take them on.

(a) Any young cockerel or pullet with dark leg colour should be culled.

(b) Young pullets heavily covered in black feather early on will generally clear as they reach adulthood. Any showing 'mealiness' (white intermixed with red) on the breast or peppering (black spots on red) on the back should be culled.

(c) In the young males, look at the two distinct lines of adult feathering growing down either side of the breast and if there is a lot of black showing in the feathers, cull them. Likewise any birds showing ginger in the neck and saddle hackles or red in the under colour should be culled.

The Sussex breed can be regarded as one of the grand large fowls if bred to Standard and there are sufficient varieties within the breed to present a breeding challenge for any keen fancier. With time and personal attention, their quiet docile nature brings them very close to their keeper.

Welsummer

Breed History

The Welsummer is an attractive partridge coloured fowl which first came to general attention because of the large brown eggs some strains lay. Other strains lay mottled eggs, that is, tinted eggs with brown spots or speckles on them. In Germany they have become very popular where particular emphasis has been placed on their egg laying ability. The Welsummer is classed as a light breed.

Welsummers take their name from the village Welsum in the Netherlands around which they were developed in the early 1900s. The original stock was of Partridge Cochin and Partridge Wyandotte breeding but had later infusions of Barnevelder, Rhode Island Red and Partridge Leghorn blood used on selected stock. The breed was first exhibited at the First World Poultry Congress held at The Hague (Netherlands) in 1921. The breed was imported into Great Britain in 1928 after reports that the Welsummer matured quicker than the Barnevelder but still laid the desired dark brown eggs so eagerly sought by the British.

The Partridge variety was standardised in 1930 with a Silver Duckwing variety accepted some years later. Bantams are now standardised as well. The British Standard is interesting because it directs breeders and judges to assess the productive features of the breed as well as its natural beauty. For this reason, the breed should appeal to those fanciers who are interested in a utility type fowl which is not white, black or red.

Welsummers reached Australia in the late 1920s and were used in egg laying trials, but their productive record could not match that of the Australian developed White Leghorn, Australorp or Langshan. It was also noted that the egg colour of the last few eggs of a clutch (eggs laid on successive days) tended to be lighter than those earlier in the clutch. It is a breed that should be enjoying greater support, particularly in the hobby farming belt.

Positive Features to Look For *(colour plates p.94)*

These are listed as follows:

1. Carriage: Alert and active.
2. Type: Balanced 'U' shape across back, top of the tail level with the head.
3. Back: Broad and straight.
5. Wings: Medium length, tucked well up against the body.
6. Tail: In the male the size should balance the bird; the sickles, tail, side hangers should form an 'all in one' appearance.
   In the female rises from the back at about 45°.
7. Saddle: Full enough to give a gentle curve to the backline.
8. Body: Long, wide, deep (back to the end of keel).

9. Head: Small, fine.
10. Beak: Short, strong—yellow or horn.
11. Eye: Bold, full, keen and red in colour.
12. Comb: Single, medium size, firm, upright with 5 to 7 serrations in the male, fine texture.
   In the male the rear part not to touch the skull or neck.
13. Face: Smooth, open.
14. Wattles: Medium size.
15. Ear lobes: Red.
16. Neck: Fairly long and upright. In the male, should account for one third of the height when viewed from the side.
17. Legs and thighs: Should be just less than one third of the height when viewed from the side, legs rich yellow colour.
18. Toes: Well spread.
19. Plumage: In the male, tight, shiny and silky.
   In the female, soft and silky in appearance.

Handling is an important feature with this breed. In general, the birds should 'look small but handle big'. The male needs to feel compact, firm, have a curve up the breast and not feel angular. The female should feel like a 'business-like layer' with plenty of capacity in the abdomen to house the essential egg laying organs.

Negative Features to Avoid

Features to avoid are those associated with non-productive fowls and colour faults associated in general with Partridge coloured fowls. They are listed as follows:

1. Sickles widely spaced from the tail.
2. Grey or green eyes.
3. Twisted or wrinkled combs.
4. 'Fly away' combs.
5. Coarse faces.
6. Beetle brows.
7. Coarse boned birds.
8. Overweight or underweight birds.
9. White in red lobes.
10. Brown stains or patches on legs.
11. Narrow width of feather in the female.
12. Baggy feathering around the thighs.
13. Visible black striping in the neck hackle of the male.
14. Birds with white flight or tail feathers.
15. Males with white in the base of the tail.
16. Hens with light coloured breasts, for example salmon pink.
17. Females with pronounced breast shafting.
18. Females with laced feathers on the back.
19. Females with 'foxiness' on the wings.

Breeding Hints

1. When looking at the scale of points allocated, the breeding priorities could be ordered as follows. Firstly, handling, size and productiveness. Secondly, type and colour. These two areas make up 70 out of the 100 points allocated.

2. Since there are not a lot of Welsummers about in this country, it would be important to develop two strains within a stud. That way, when it is time for an outcross, blood of a close line is readily available.

4. Egg colour, mottling and texture can all be selected for in a breeding program. These features can be further developed by a carefully planned line breeding program. The key is to use sons of hens that produced the desired egg features as the hens will pass the desired features on to their granddaughters through their sons. For keen breeders, it is possible over a number of years to breed in dark brown egg laying ability into exhibition stock, but it requires dedication and perseverance.

4. Exhibition males and females can be bred from the one breeding pen if due care is exercised in selecting it, as the body colour of the female is a warm brown, different from the body colour of Modern Game, Old English Game and Brown Leghorn. A golden shaft is a feature of the female neck hackle. There could be a temptation to breed from 'foxy winged' females to produce bright exhibition males, but this is the first step down the road of double mating, something that is not necessary in this breed.

Welsummers are a breed that should be enjoying a much greater following than they do as they offer bright colour and dark brown eggs in the one breed without the need to double mate.

Wyandotte

Breed History

Wyandottes originated in New York State of the USA, and were given the name of a North American tribe of Indians who used to live in the area. The breed was developed from Chittagong, Malay, Cochin, Dorking and Brahma blood. The first of many members of the Wyandotte family, the Silver Laced Wyandotte was standardised in 1883 and admitted to the American Standard of Perfection in the same year. Many of the other colours followed soon after.

The breed soon became accepted as a utility fowl for both eggs and meat in both Great Britain and the USA. In the 1920s, particularly in Great Britain, the breed was subjected to the ravages of commercial exploitation, being simply mass produced with little regard for the Standard. As a result, problems arose with small egg size, poor laying and a decline in fertility, often as low as 50%. As a commercial proposition, the breed fell into disfavour.

It was during this time that the Wyandotte was crossed with other breeds to improve its commercial features. It was found that the White Wyandotte showed sex linkage when crossed with the Brown Leghorn and the Rhode Island Red. The Leghorn blood also lengthened the body and, while improving the egg laying ability of the fowls, destroyed the true Wyandotte type. Such was the change in type, that two distinct forms of Wyandotte appeared: the exhibition or Standard Wyandotte and the Utility Wyandotte. The extremes of both types and their followers did much to ruin the commercial features of the breed. As a result, the more commercially vigorous White Leghorn, Rhode Island Red and Australorp replaced the Wyandotte as a commercial fowl.

Selected strains of Wyandottes performed well in early laying trials conducted in Australia but soon found it difficult to compete with the then developing strains of Langshans, Australorps and White Leghorns.

Today, the Wyandotte breed is regarded as a fancier's breed, with the Wyandotte Club of Australia recognising the following colours: White, Black, Columbian, Gold Laced, Partridge, Silver Laced, Blue, Barred, Cuckoo, Buff Columbian and Crele. Mention has also been made of Blue Laced, Buff Laced and Birchen in poultry literature. Bantams exist in most of the above mentioned colours. The Wyandotte, with its distinctively curvaceous outline and wide range of colours, is a breed with wide appeal.

Positive Features to Look For *(colour plates p.95)*

. The Wyandotte is regarded as a 'bird of curves' and an important point to look for is the 'symmetry of the whole bird'. In days gone by, the common way of expressing the desired shape was by likening it to a 'U'.

The bird should have a distinctly 'U' shape on the underline and the top line. This did not indicate the true shape of the topline as can be seen in the accompanying type diagrams. Besides, when talking in terms of letters, a two dimensional impression is imparted instead of the desired three dimensional form of the fowl.

2. From the side view, the fowl should look as follows:

(a) Body: Round, medium length and plenty of depth.
(b) Neck: Short and curved.
(c) Breast: Round, broad and deep. Breast and vent fluff should form a semi-circle. The breastbone level with the hocks.
(d) Back: Medium length but wide along its length.
(e) Wings: Well tucked up beside the body.
(f) Saddle: Wide, well filled out and rising in a concave sweep to the tail.
(g) Fluff: Moderately full.
(h) Tail: Carried a little lower than the head.

3. When the fowl is viewed from the front or rear, the following should be evident:

(a) The bird should be broad and well rounded, with medium length legs.
(b) The legs themselves should be well shaped and yellow in colour.
(c) The sides should be well rounded.

4. Turning our attention to specific head features, the following are what to look for:

(a) Eyes: Bold, round and reddish bay in colour.
(b) Beak: Medium length and yellow in colour.
(c) Face: Smooth and of fine texture.
(d) Comb: Rose comb that is firmly placed on the head. The comb's surface should be oval in shape with the surface covered with small round points called 'workings'. The leader (rear of comb) needs to go to a well defined point and follow the shape of the skull.
(e) Wattles: Medium length, well rounded, uniform and smooth.
(f) Ear lobes: About a third the length of the wattles in size and oblong in shape.
(g) Face, lobes, wattles and comb should be bright red.

Negative Features to Avoid

Starting with a general comment on the overall impression of the fowl, reject those that are out of general balance and symmetry. Now turning our attention to the various parts of the fowl, avoid the following:

1. Body: Fowls that are angular instead of curved. Likewise, those with excessive size, narrow bodies and those with bodies carried too high on their legs.

2. Underline fluff and feathers: Any that are too lon and dense in this area as this leads to the fow having no underline. The hocks and shanks ar also covered. The Wyandotte should not b Cochin or Orpington in appearance in this area
3. Front and breast: Cull out any that cut awa towards the thighs. These fowls do not have th full round breast required.
4. Neck: Partly arched or curved necks.
5. Back: Long, flat, narrow backs.
6. Cushion: Balled cushioned birds, particularl females. Also those where the cushion finishe over the tail.
7. Tail: Narrow 'pinched' tails behind well develope cushions. Any fowl showing a lack of tail.
8. Wings: Wings that are too long as this spoils th symmetry of the bird. Split wing. Any with wing carried too low.
9. Legs: Square shanks and those with feathere shanks.
10. Sides: Flat sided birds.
11. Comb: Lack of 'workings' on combs. Thos whose combs do not sit squarely or firmly on th top of the skull.
12. Beak: Excessive length of beak.

Breeding Hints

Mention should be made of the fact that th Wyandotte–Silkie Cross has long been recognised a one of the best broody hens for raising chickens. Th Wyandotte reduces the problem of the long stream feathers of the Silkie strangling young chickens, whil imparting a volume of feathers that ensures the he can raise a large number of chickens.

Before turning to specific colours, some genera remarks need to be made. Firstly, attention must b paid to the influence of the female on type in th offspring. The following is the general type needed She should have a *short* broad back, body, neck an tail plus a *deep* round body with a broad, deep, roun breast and low keel set. Secondly, turning our attentio to the comb, a female with a slightly larger comb wit plenty of 'workings' will produce better cockere combs, however this could lead to a 'double mating situation if taken too far.

1. White
Whites first appeared as sports from the Silver Lace variety. They have the advantage that they do no require double mating.

However, there are problems with maintaining pure white colour as there is a tendency for sandines or brassiness to appear across the shoulders. I

White Wyandotte bantam male. 1988 Sydney Royal Easter Show. Owner: Frank Catt.

White Wyandotte bantam female. Bantam Club of N.S.W. Annual Show 1988. Owner: F. Catt.

White Wyandotte large fowl female. 1988 Sydney Royal Easter Show. Owner: Alex Lamont.

Columbian Wyandotte bantam male. Bantam Club of N.S.W. Annual Show 1988. Owner: L. Sinclair.

Columbian Wyandotte bantam female. Bantam Club of N.S.W. Annual Show 1988. Owner: L. Sinclair.

Black Wyandotte bantam female. Bantam Club of N.S.W. Annual Show 1988. Owner: J. Clark.

Black Wyandotte bantam male. Bantam Club of N.S.W. Annual Show 1988. Owner: J. Clark.

addition, there are those with a permanent 'sappy' (yellowish tinge) to the white colour. In Wyandotte terms, the desired white is known as a 'stay white' strain and this is the colour type to be sought.

As with most white fowls, Whites must be sheltered from the weather or the plumage spoils and tends to discolour.

The 'stay white' trait appears to be a sex linked characteristic. It needs to be traced in the strain, identified in specific fowls and bred into future lines.

If a fowl has a few smutty feathers across the back and is known to come from a stay-white strain, keep it for breeding as the desired feature will be passed on to the offspring.

2. Black

As with most fowls with black plumage and yellow legs, double mating is required to breed exhibition males and females.

(a) For the cockerel breeding pen a male with yellow legs, green sheen and undercolour as dark as possible needs to be mated to a female from a cockerel breeding line with some green sheen showing, and dark legs with yellow showing through on the soles and the webs of the feet.

(b) In the pullet breeding pen, an exhibition female with good yellow leg colour needs to be mated with a male from a known pullet breeding strain with true yellow legs, a little red in the hackles and beetle green top colour with white in the undercolour.

3. Partridge

The unfortunate position with the Partridge variety is that in the past emphasis has been placed on colour pattern at the expense of type and egg laying ability. The sequel to this today is that the egg laying ability of the female line could best be described as 'dodgey'. Past breeding efforts can be summed up in the following often quoted statement: 'The cock is a thing of beauty; the hen a work of art'.

To successfully breed Partridge Wyandottes, a double mating system has to be used.

(a) The cockerel breeding pen requires an exhibition male with a solid black breast and clean top colour. Be on the look out for the following faults. Firstly, black stripes that run through the gold lacing on the hackle. Secondly, black edging around the gold lacing on the hackles. Thirdly, shaftiness or straw coloured quills. Fourthly, light undercolour which generally shows first on the sickles and tail feathers. The female should come from the cockerel breeding strain and be of excellent Wyandotte type. It is important that her colour points do not clash with those of the male.

Attention needs to be paid to the following. Firstly, the pencilling should be indistinct. Secondly, the neck hackle should show a solid stripe in each feather. Thirdly, the hackle should show a bright outer lace and no shaft. Fourthly, a sound black tail. Finally, there is much debate as to the best body colour in the female, some going for a darker partridge shade whilst others prefer the washed out colour.

(b) In the pullet breeding pen, the opposite applies in principle. The female should be an exhibition Wyandotte type with attention paid to the following. Firstly, there should be three distinct markings of black on an oak leaf brown ground colour. Secondly, the markings or pencilling should be finely and sharply drawn, not broken. Thirdly, the undercolour fluff should be grey. Fourthly, watch out for weak markings on the breast. The pullet breeding male, on the other hand, should be from a pullet breeding line and be weak in the female strong points so there is no clash. Other helpful points to look for are as follows. Firstly, weak saddle and neck hackle striping. Secondly, a broken front with brown lacing or splashes on the breast. Thirdly, look for female markings on the back. These often require close examination to find. Fourthly, a bright top colour. Fifthly, a small comb.

It is important not to cull Partridge Wyandottes too early on colour. It is six to seven months before the true adult plumage shows through and a judgement can be made.

4. Silver Pencilled

These are similar to the Partridge variety except that the ground colour is different. The remarks made about the Partridge variety apply equally to the Silver Pencilled variety. The ground colour of the female is a soft steel grey instead of the oak-leaf brown of the Partridge female. A double mating system is also required.

5. Columbian

There are some doubts over the origin of the Columbian variety, with some saying that it is a product of a White Wyandotte—Barred Rock cross, whilst others feel that it resulted from a White Wyandotte—Light Brahma cross.

To breed the Columbian pattern successfully requires a 'balanced mating' approach, not to be confused with double mating, although the end result of the balanced mating may well be to produce exhibition cockerels or pullets. The Columbian pattern will breed true over a number of seasons but tends to 'wash out' in the dark points, thus needing some corrective action in the breeding pen. To breed this pattern, it may be necessary to maintain a few birds for this purpose. Of course, they are not suitable for exhibition.

Now turning our attention to some specific points:

(a) *Body colour*
The back needs to be pure white in the web. The black points should be jet black with green sheen and not grey or brown.

(b) *Hackle*
To produce good hackle in the male offspring, the female needs to have sound stripes and dark undercolour in the neck hackle. Another way to overcome light hackle is to use a very dark hackled male or female laced well up the head, even if there is an excess of black in the body. Such a bird is an example of a stock bird that needs to be kept at home and neither shown nor culled.

(c) *Undercolour*
Blue or slate undercolour is an indicator of the amount of pigment (black) carried by a bird. Slaty undercolour can be used to counter the creamy or brassy tinge that occurs in top colour in some fowls. Likewise, a lighter undercoloured bird can be used to dilute an excess of black in heavily pigmented birds.

(d) *Wings*
Try to balance wing colour with other black points. It is easier to get good wing colour in the male than the female. This is why it is important to retain a hen with good wing colour, as she is very useful as a breeder.

(e) *Tail*
A solid black tail in the female is needed to produce a solid black tail in the male offspring.

(f) *Saddle*
Some striping or 'ticking' should be present in the male. This also indicates a strength of pigment.

As can be seen from the foregoing, breeding the Columbian variety is a balancing or 'juggling' process if the true pattern is to be produced in both males and females with any consistency. It also necessitates having

a reserve of breeding stock on hand. Therefore, this breed is best suited to a specialist rather than a person prepared to have only one breeding pen.

6. Silver Laced
Double mating is essential if exhibition males and females are to be bred. Both strains must be kept separated.

(a) The cockerel breeding pen would be made up of an exhibition male with a sound black tail, breast free of double lacing, a distinct black stripe down the neck and saddle hackle, and a silvery white head, back and wingbows. His partner would be a female from a known cockerel breeding strain with a well laced breast free of irregular lacing, a sound black tail and 'mossy' or 'peppered' back and cushion. Avoid any female that is sooty or brassy in the neck hackles. Sometimes cockerel breeding females come brassy winged.

Look carefully at the undercolour as there is a connection between light fluff, peppery thighs and the two faults, double lacing and horse shoe lacing.

(b) The pullet breeding pen would be headed by an exhibition female from a pullet breeding strain with a clear centre to each feather, free of ticking or mossiness, and a fine lacing around each feather. The male needs to be from a known pullet breeding strain and possess a clear, sound laced breast plus some white in the tail.

Watch the lacing thickness as this can be balanced by judicious mating.

7. Gold Laced
The Gold Laced variety is the same as the Silver Laced except for ground colour. The same breeding hints would apply.

The Wyandotte family offers a sufficiently wide range of colours and breeding challenges to satisfy even the most specialist of poultry fanciers. This is probably why the breed still has a wide following across this country.

Large Silver Laced Wyandotte female. Fairfield Poultry Club Autumn Show 1988. Owner: J. Gardner.

Large Silver Laced Wyandotte male. Fairfield Poultry Club Autumn Show 1988. Owner: J. Gardner.

Faults *(colour plates on p.96)*

Fault: 1. "Goose winged". Wing ends carried across the back in front of the tail.
2. "Shaftiness" also seen in feathers.

Fault: A bad case of "shaft" or "shaftiness" in the feather.

Fault: White in black tail feathers of a Light Sussex male.

Faults: Dropped tailed bird. Slipped wing.

Faults: 1. Slipped wing. 2. Overshot comb over beak.
3. Excessive comb.

Faults: Split wing. Gap between primary and secondary wing feathers.

Faults: 1. Squirrel tail. 2. Slipped wing. 3. Frizzled neck hackle. 4. Lopped comb.

Faults: 1. Undershot beak. 2. Double serration at rear of comb.

Fault: Double serration in the centre of the comb.

Fault: Double serration of rear comb serration.

Fault: Small serrations in comb above beak. Should be free of these.

Fault: Twisted blade at rear of comb.

Faults: 1. Badly "thumb marked" in front of comb. 2. Crease in comb above the beak.

Dubbed White Leghorn male.
Faults: 1. Blisted lobe. 2. Undershot beak.

Faults: 1. Folded lobe. 2. Red in white lobe. 3. Comb double folded over beak.

Fault: Comb double folded over beak.

Faults: 1. Comb twisted over beak.
2. Comb not straight over the top of the skull.

Faults: 1. Overhanging eyebrows or "beetle browed".
2. Comb twisted over beak. 3. Badly creased comb over skull. 4. Uneven wattles.

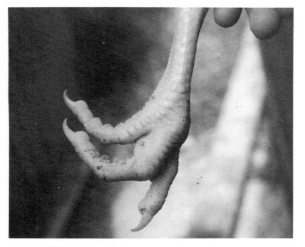

Fault: "Crooked toes". Central and outer toe badly bent.

Fault: Bad case of scaley leg mite.

Fault: Feather stubs between toes of a Light Sussex.

Fault: Feather stubbs in the shanks of an Australorp.

Solving Problems and Chasing Information

This section of the book is an attempt both to help the fancier solve various problems and also to provide resource information which may help in their poultry endeavours. Being a fancier means that you have to be aware of a wide range of sources, sometimes with no immediate and direct relationship with poultry.

If you wish to be a successful information-gatherer and problem-solver, you must develop the skill of 'networking'. If the person you initially contact cannot give you the answer you seek, do not be frightened to ask for a possible lead. It is surprising what information is out there if you are persistent.

Please note that the advice given and organisations mentioned by the author are not meant to imply that a recommendation or guarantee is being given, but rather that a potential source of help is being suggested. Dealings with any of these sources is up to the individual.

Sources of Information

Books

Whilst specialist poultry books are not regularly shelved in all municipal libraries, most of these libraries have extensive links with other libraries. It is a matter of asking library staff what library networks they belong to and requesting a search for specific titles. Those familiar with computer catalogues can do this themselves. If you are not skilled in this area, then ask the library staff to help.

With the advent of the Internet, it is possible to do a library search for a title on a worldwide basis. You need to contact the library with regards to their lending or copying policy. Many will send photocopies of relevant parts of a book. This has dramatically increased the fancier's access to information on a global basis.

For those who favour historical aspects of poultry, the Mitchell Library in Sydney has a complete bound set of the *Poultry* newspaper dating back to the first edition! The bound copies cannot be borrowed but can be read at the library. One year's copies are available at a time.

There are a number of organisations that do carry specialist poultry books such as the Bantam Club of N.S.W. and the New England Exhibition Poultry Association. Book lists can be obtained by contacting those organisations.

Magazines

The *Australasian Poultry Magazine* has emerged in recent years as the premier poultry publication. It is a bimonthly, providing an authoritative source of information. It is published by the:

Poultry Information Publishers
P.O. Box 198
Werribee
Vic. 3030

The New England Exhibition Poultry Association has two publications of use to fanciers. The first is a monthly magazine *Chicken Chat* and the second is the *Year Book* containing a listing of 200 clubs in Australia, their addresses and show dates. The *Year Book* also contains judges Australia-wide, and gives details of over 2000 breeders throughout Australia as well as the major show results for the previous year.

This organisation can be contacted through:

New England Exhibition Poultry Association
P.O. Box 284
Glen Innes
N.S.W. 2170

Poultry Feed

Most large poultry feed companies are vertically integrated into poultry production. As a consequence, they have considerable technical information on nutritional aspects. A check of phone book and contact can yield useful information.

Poultry Equipment

Multiquip
P.0. Box 4
Austral
N.S.W. 2171

This company stocks a wide range of incubators, brooders, fertility testers, feeders, waterers and leg rings.

Bellsouth Pty Ltd
P.O. Box 1233
Narre Warren
Vic. 3805

This company offers a mail order service on a wide range of poultry needs such as incubators, feeders, drinkers, medications and disinfectants. It also carries a range of books.

I.J. Trass & Sons
Ph (074) 963 686

This company manufactures poultry show cage fronts and complete collapsible show pen fronts. They will freight these anywhere in Australia. To be competitive, your fowls will have to be trained, so show pens need to be part of your yard's structures.

Diseases

NSW Agriculture and other State Departments of Agriculture publish *Agfacts* or *Agnotes* on the major poultry diseases. Whilst these are aimed at the commercial producer, important background information can be gleaned from them. The facts or notes are arranged under the Agdex system so information can be accessed across Australia. Most State departments have specialist poultry vets who can be phoned for advice but do not make individual calls on people unless your fowls have some rare and exotic disease!

Webster's vaccines now operate through a series of rural retailers but do publish useful technical information on diseases and their range of vaccines. The problem for most fanciers is that the smallest vaccine packs are too large for the individual and hence the need to club together with others to coordinate the vaccination program to cut down on waste.

Chemicals

Vetafarm Pty Ltd
P.O. Box 5244
Wagga Wagga
N.S.W. 2650

This company carries stocks of specialist disease control chemicals and nutritional supplements.

Specialist Hatcheries

A number of specialist pure breed hatcheries have been established to supply pure breed day-old chickens. Whilst the best way to start into a breed is to buy breeding stock from a reputable breeder, day-olds could provide an alternative, provided you are given a background to the stock producing the chickens. Such specialist hatcheries include:

Darling Downs Hatchery
P.O. Box 176
Toowoomba
Qld 4350

Fancy Feathers Farm
R.M.B. 318
Mardi Rd
Wyong
N.S.W. 2259

Greg Nordstrom Poultry
P.O. Box 722
Armidale
N.S.W. 2350

Brianne Park
Back Callington Road
Callington,
S.A. 5254

Index